Tsukuba International Conference on Materials Science 2013

IOP Conference Series: Materials Science and Engineering Volume 54

Tsukuba, Japan
28 August – 6 September 2013

ISBN: 978-1-63439-865-7
ISSN: 1757-8981

Printed from e-media with permission by:

Curran Associates, Inc.
57 Morehouse Lane
Red Hook, NY 12571

Some format issues inherent in the e-media version may also appear in this print version.

Printed by Curran Associates, Inc. (2015)

Institute of Physics
Dirac House, Temple Back
Bristol BS1 6BE UK

Phone: 44 1 17 929 7481
Fax: 44 1 17 920 0979

techtracking@iop.org

Additional copies of this publication are available from:

Curran Associates, Inc.
57 Morehouse Lane
Red Hook, NY 12571 USA
Phone: 845-758-0400
Fax: 845-758-2634
Email: curran@proceedings.com
Web: www.proceedings.com

TABLE OF CONTENTS

IOP Conf. Series: Materials Science and Engineering **54** (2014) 011001 doi:10.1088/1757-899X/54/1/011001

Preface

Tsukuba International Conference on Materials Science (TICMS) was held from 28th August to 6th September, 2013 for the celebration of 40th year anniversary of the University of Tsukuba. The conference was organized by the Division of Materials Science, in cooperation with the Graduate School of Pure and Applied Sciences, and Tsukuba Research Center for Interdisciplinary Materials Science.

The purpose of the conference was to provide a unique forum for researchers and students working in various fields of materials science, which have been progressing so rapidly that no single society could cover. The conference consists of following seven workshops to cover various fields. The organizing committee believed that the conference gave all participants new insights into the widespread development of materials science and enhanced the circulation, among them, of information released at the conference. The organizers are grateful for the financial support from University of Tsukuba.

This volume contains 25 selected papers from invited and contributed papers, all of which have been screened on the basis of the standard review process of the program committee.

The editors express their thanks to those authors who contributed the papers published in this proceedings, which reflects the scientific value of the conference.

Nov. 20, 2013

Seiji Kojima, Prof. Dr.
Chair, Division of Materials Science
Chair, Doctoral Program in Materials Science

TICMS 2013 (http://www.ticonfms.tsukuba.ac.jp/)
Workshop list

(i) The 13th Japan-Korea Joint Workshop on Materials Science
(ii) Summer School of Biomaterials Science
(iii) The Japan-Korea Joint Workshop on Shape Memory and Superelastic Technologies
(iv) The 2nd Workshop on THz Radiation from Intrinsic Josephson Junctions
(v) The 3rd German-Japan Nanoworkshop
(vi) TICMS and IWP Joint Workshop on Conjugated Polymers
(vii) International Workshop on Science and Patents (IWP) 2013

IOP Conf. Series: Materials Science and Engineering **54** (2014) 011002 doi:10.1088/1757-899X/54/1/011002

Peer review statement

All papers published in this volume of *IOP Conference Series: Materials Science and Engineering* have been peer reviewed through processes administered by the proceedings Editors. Reviews were conducted by expert referees to the professional and scientific standards expected of a proceedings journal published by IOP Publishing.

IOP Conf. Series: Materials Science and Engineering **54** (2014) 012001 doi:10.1088/1757-899X/54/1/012001

Broadband terahertz time-domain spectroscopy : crystalline and glassy drug materials

Seiji Kojima, Tomohiko Shibata, Hikaru Igawa, and Tatsuya Mori
Division of Materials Science, University of Tsukuba, Tsukuba, Ibaraki 305-8573, Japan

E-mail : kojima@ims.tsukuba.ac.jp

Abstract. Low-energy IR active modes of glassy and crystalline drug materials were studied by the broadband Terahertz Time Domain Spectroscopy (THz-TDS) in the frequency range from 0.5 to 6.5 THz using a Cherenkov type THz generator. In order to determine the real and imaginary parts of complex dielectric constant, all samples were measured by the transmission using a pure pellet without mixing polyethylene. For glassy indomethacine, the broadband THz spectrum of real part of dielectric constant shows step-wise decrease with the increase of frequency, while the imaginary part shows a broad peak at about 3 THz reflecting quenched glassy disordered structure. The observed spectra of crystalline racemic ketoprofen show the noncoincidence of peak frequencies between low-frequency Raman scattering and THz absorbance spectra. It can be attributed to the fact that the mutual exclusion principle between Raman and IR activities holds below 6 THz.

1. Introduction

Vibrational spectroscopy is a powerful tool to elucidate not only the static crystal or molecular structure but also the dynamical properties related to bond strength and molecular interaction, etc. The far-IR spectroscopy has been extensively applied to various kinds of crystalline and glassy materials to investigate low-energy excitations such as ferroelectric soft modes, polaritons, excitons, boson peaks, etc., while most far-IR studies reported results on the frequency-dependent absorbance or transmittance, and the real and imaginary parts of dielectric constants were not reported.

The new technique of the coherent terahertz generation using a femtosecond pulse laser enables the unique determination of a complex dielectric constant without the Kramers-Kronig transformation, and terahertz time domain spectroscopy (THz-TDS) has attracted much attention. However, the most of the THz-TDS works until now was done below 2 THz, which is equivalent to 67 cm^{-1} [1]. Very recently widely tunable monochromatic Cherenkov phase-matched THz wave generator was developed using nonlinear optic crystals of ferroelectric MgO doped $LiNbO_3$ [2]. The frequency range of THz-TDS has been extended up to 6.5 THz by such a broadband THz generator. In addition, the progress of the rapid scan method enables to acquire high quality THz spectra by the use of high-speed asynchronous optical sampling (AOS) technique [3].

In this work, transmission broadband THz spectra in the frequency range from 0.5 to 6.5 THz were studied in glassy and crystalline states of polymorphic and polyamorphic drug materials by using a THz-TDS equipment with a Cherenkov type THz generator and the high-speed AOS technique. The

THz spectrum was compared with the low-frequency Raman spectrum to discuss the exclusion principle of selection rules between IR and Raman scattering spectroscopy.

2. Experimental

Indomthacine with the purity ≥ 99% and racemic ketoprofen with the purity ≥ 98% were purchased from Sigma-Aldrich. The glassy state of indomethacine was obtained without further purification by melt-quenching at 80 K. All spectra were measured using a thin plate or a pressed pure pellet without mixing other solid powders such as polyethylene to determine real and imaginary part of dielectric constant accurately.

THz transmission spectra were measured in the frequency range from 0.5 to 6.5 THz using a THz-TDS equipment (TAS7500SU, Advantest Co.) with a Cherenkov type THz generator and the high-speed AOS technique. The low-frequency Raman scattering spectra were measured by a single frequency green-YAG laser with wavelength 532 nm and a triple-grating monochromator (T64000, Horiba-Jobin-Yvon) with additive dispersion [4]. The comparison of THz radiation power between a conventional low-temperature grown GaAs dipole photoconductive antenna and a Cherenkov type THz generator is shown in figure 1. The photoconductive antenna has a broad peak at about 0.7 THz, while the peak frequency of a Chrenkov type is around 2.5 THz and very broad. Therefore, the latter type generator enables us to observe a broadband spectrum from 0.2 to 6.5 THz.

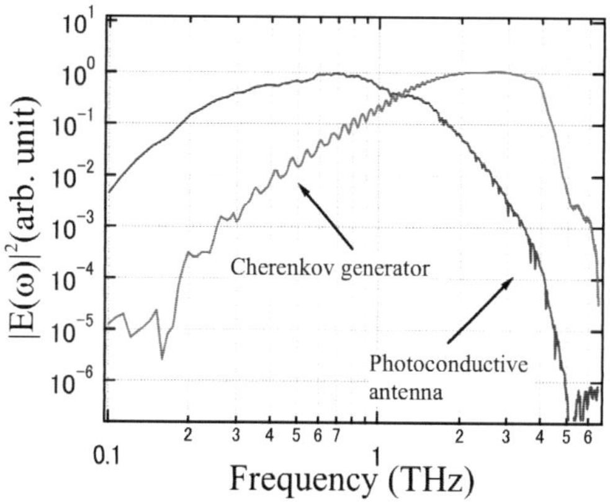

Figure 1. Comparison of THz power spectra between a Cherenkov type THz generator and a dipole photoconductive antenna.

3. Results and discussion

The most of drug materials are polymorphic and polyamorphic originated from multi-basin structure in the potential energy landscape. The energy of these basins is nearly degenerate, and the transition among basins can be easily occurred by the change of temperature, pressure, and other chemical conditions. The dynamical properties in the THz frequency range are very sensitive for the difference among polymorphic and polyamorphic states. Therefore, the investigation on each state is very important to give new insights into polymorphic and polyamorphic nature.

3.1. Complex dielectric constant of glassy indomethacine

The indomethacine, 1-[19]-5-methoxy-2-methylindole-3-acetic acid with chemical formula $C_{19}H_{16}ClNO_4$, is a nonsteroidal anti-inflammatory drug (NSAID) used for the treatment of pain. Molecular structure of indomethacine is shown in figure 2(a). Up to the present several papers reported THz absorbance spectra of glassy and crystalline indometacine [5-7]. In refs. 5 and 6, THz absorbance spectra of glassy indomethacine were reported below 75 cm^{-1}~ 2.2 THz. The gradual increase of absorbance with the increase of frequency was observed, while no peak was observed in glassy samples. Very recently the high frequency limit was extended up to 6 THz by the use of a GaP THz signal generator [7]. The broad peak near 3 THz was reported in the absorbance spectrum of glassy indomethacine using the mixed pellet with polyethylene, while the complex dielectric constant was not reported.

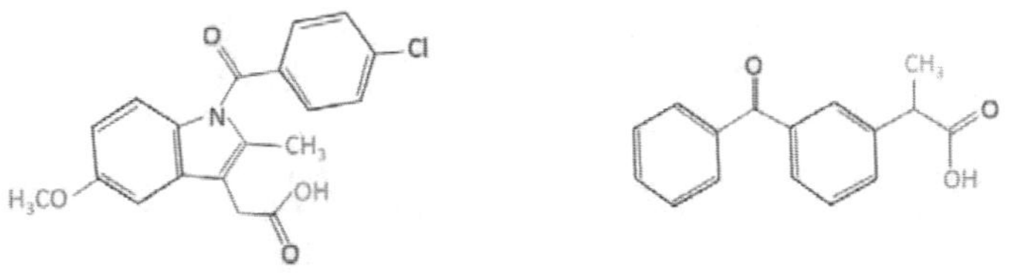

(a) (b)

Figure 2. Molecular structures of (a) indomethacine and (b) ketoprofen.

In the present study, the thin plate of glassy indomethacine was prepared by melt-quenching method in liquid nitrogen. Since polyamorphic nature was reported for indomethacine, and the present glass was prepared from the γ−form crystalline powders [5]. In contrast to previous THz studies, the pellet mixed with polyethylene powder was not used. Therefore, the determination of real and imaginary parts of dielectric constant is possible by the transmission measurement. The real and imaginary parts of complex dielectric constant spectra were observed as shown in figure 3 in the frequency range from 0.5 to 6.5 THz. The real part of dielectric constant shows the step-wise decrease with the increase of frequency and the broad and asymmetric peak of the imaginary part near ν_1=3 THz.

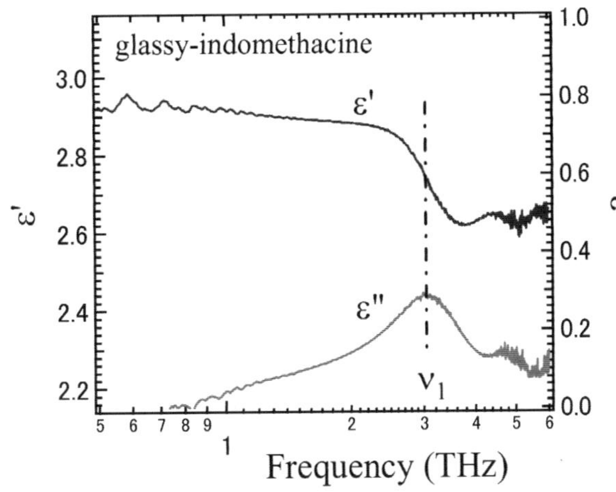

Figure 3. Real and imaginary parts of THz complex dielectric constant of glassy-indomethacine melt quenched from the γ-form.

IOP Conf. Series: Materials Science and Engineering **54** (2014) 012001 doi:10.1088/1757-899X/54/1/012001

As the common nature of the low-energy dynamics of glassy materials, a boson peak has been observed as a broad and asymmetric peak at few THz by Raman scattering, far-infrared spectroscopy, neutron and X-ray inelastic scattering [4, 8]. It reflects the vibrational density of states (VDOS) and the broad peak has been also observed in the temperature dependence of C_p/T^3 curve at low temperatures, where C_p is the heat capacity. It has been known that the boson peak frequency has the strong correlation with the shear modulus [8], and the molecular dynamical simulation predicted recently that the origin of the boson peak is the Ioffe-Regal limit of transverse acoustic mode, and its frequency is proportional to shear modulus [9].

The broad and asymmetric features of the peak of imaginary part near 3 THz looks like a boson peak, while considering the small shear modulus of organic glasses, the boson peak frequency can be less than 1 THz. Very recent study of glassy indomethacine by inelastic X-ray scattering reported that the Ioffe-Regal limit is about 0.6 THz [10]. As a general trend, a strong glass shows an intense boson peak and a weak fast relaxation process in an inelastic scattering spectrum, while a fragile one shows a weak boson peak and an intense fast relaxation process [11]. Since the large value of fragility index of indomethacine, m=60-79, was reported, the weak intensity is expected for a boson peak. Probably it is the reason why we cannot observe the boson peak in the imaginary part of dielectric constant in the present experiment. Considering the fact that the broad peak at 3 THz does not appear in the spectra of a crystalline state, it can be attributed to the some kind of vibrational mode with the distribution of mode frequency reflecting the distribution of bond length and bond angle in glassy structure and the high damping factor caused by the quenched glassy disordered structure. One possible origin of this peak is the vibration of the hydrogen bonding between molecules.

3.2. Comparison between Raman and THz spectra of crystalline racemic ketoprofen

Racemic ketoprofen, (RS)-2-(3-benzoylphenyl) propanoic acid with chemical formula $C_{16}H_{14}O_3$ is a nonsteroidal anti-inflammatory drug (NSAID) used for the treatment of inflammation and pain. Molecular structure of ketoprofen is shown in figure 2(b). It contains a chiral center at the α-carbon to the carboxyl function and R(−) and S(+) enantiomeric forms are available [12]. Ketoprofen is polymorphic, and nine conformers were known [13]. The Raman scattering spectrum of the present sample is assigned to "conformer A" according to the experimental and calculated mode frequencies reported in ref. 13. A THz spectrum was measured using a pure pellet without mixing polyethylene.

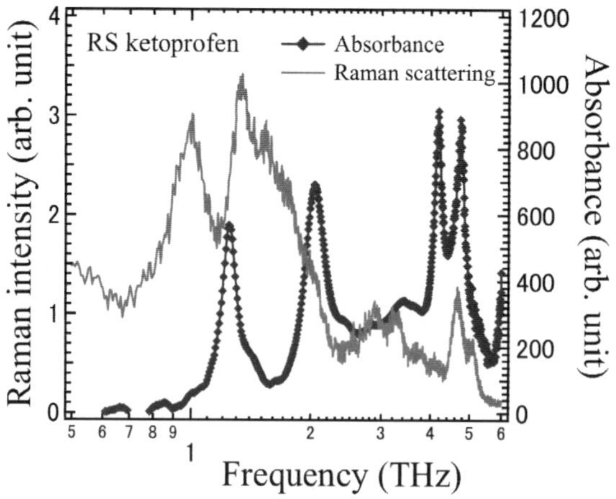

Figure 4. Raman scattering and THz-TDS absorbance spectra of crystalline racemic ketoprofen.

The comparison between THz-TDS and Raman scattering spectra in the frequency range from 0.5 to 6.5 THz was shown in figure 4. In a Raman spectrum, the peaks above 100 cm^{-1} such as 110, 124,138, 157, 170 cm^{-1}, etc. are in agreement with the calculated values of the conformer A reported in ref. 13 within experimental uncertainty. For the internal vibrational modes above 400 cm^{-1} (=12 THz), it was reported that the mode frequencies observed by Raman scattering and FTIR spectroscopy were in agreement within experimental uncertainty. In contrast, the external modes below 3 THz, the discrepancy of peak frequencies was clearly observed between THz-TDS and Raman scattering spectra as shown in figure 4.

When the point symmetry of a crystal has a center of symmetry, the mutual exclusion principle holds between Raman and IR activities. Some vibrational modes are only Raman active, while others are only IR active. According to the previous report, the most of mode frequencies observed by Raman and FTIR measurements are in good agreement above 400 cm^{-1} (12 THz). Therefore, the exclusive principle seems to do not hold for internal vibrational bands above 400 cm^{-1}. However, below 3 THz the exclusion principle clearly holds between Raman scattering spectroscopy and THz-TDS. Consequently, for a racemic crystalline state, the exclusive principle holds below 3 THz for external vibrational modes which include translational and rotational motions, while it may not hold above 6 THz for internal vibrational modes which are defined by the internal coordinates used to describe conformation and geometry of a molecule [13]. These facts suggest that the centrosymmetric crystal lattice with translational symmetry is formed for the center of mass of molecules, while the orientation and structure of molecule at a lattice point has some kinds of disorder in crystalline racemic ketoprofen. This kind of disorder may have some similarity to an orientational glass state such as metastable glassy crystal of ethanol (crystal-II) at low temperatures [14,15]. Orientational glass is a molecular solid in which translational symmetry coexists with the static disorder of some rotational degree of freedom.

For the characterization of polymorphic and polyamorphic drug materials, it is very important to measure both THz-TDS and low-frequency Raman scattering. The two methods have the complimentary selection rules which enable the full information on vibrational properties related to the crystalline structure, glassy intermediate range order, and their strength of intermolecular interactions.

4. Conclusion

Two non-destructive spectroscopic methods - broadband THz-TDS and low-frequency Raman scattering - were employed to elucidate the vibrational properties of glassy and crystalline drug materials. The glassy-indomethacine melt-quenched from the γ−form crystalline powder was measured by the broadband THz-TDS in the frequency range from 0.5 to 6.5 THz. The complex dielectric spectrum of glassy-indomethacine shows the broad peak of the imaginary part in the vicinity of 3 THz. It can be attributed to not a boson peak related to vibrational density of states but a diffusive normal mode with the frequency distribution and high damping reflecting the glassy disordered structure. The possible origin of this band is the weak interaction between molecules such as a vibration of hydrogen bond between molecules. In crystalline racemic ketoprofen, the difference between THz transmission and low-frequency Raman scattering spectra was clearly observed below 3 THz, and it suggests that the exclusive principle of IR and Raman selection rules holds for external vibrational modes of a present crystalline state. However, according to the experimental results in literature, the exclusive principle may not hold for internal vibrational modes above 12 THz. These facts indicate that a crystalline state of racemic ketoprofen has a crystal lattice with the center of symmetry while the structure of molecules at lattice points is disordered. It is suggested that for the study of polymorphic and polyamorphic states the measurements of both THz-TDS and low-frequency Raman scattering are important to give new insights into their structure and dynamics.

Acknowledgement
Authors are very thankful for the technical support on THz-TDS measurements to Sendai Office, Advantest Corporation.

References

[1] Kojima S, Tsumura N, Wada Takeda M and Nishizawa S 2003 *Phys. Rev. B* **67** 035102
[2] Shibuya T, Suizu K and Kawase K, 2010 Appl. Phys. Express **3** 082201
[3] Bartels A, Hudert F, Janke C, Dekorsy T and Köhler K 2006 *Appl. Phys. Lett.* **88** 041117
[4] Kojima S 1993 *Phys. Rev. B* **47** 2924
[5] Strachan C J, Rades T, Newnham D A, Gordon K C, Pepper M and Taday F 2004 *Chem. Phys. Lett.*, **390** 20
[6] Strachan C J, Taday P F, Newnham D A, Gordon K C, Zeitler J A, Pepper M and Rades T 2005 *J. Pham. Sci.* **94** 837
[7] Otsuka M, Nishizawa J, Fukura N and Sasaki T 2012 *J Infrared Milli Terahz Waves* **33** 953
[8] Kojima S, Matsuda Y, Fukawa Y, Kawashima M, Moriya Y, Yamada T, Yamamuro O and Kodama M 2010 *J. Non-Crys. Solids* **356** 2524
[9] Shintani H and Tanaka H, 2008 Nat. Mater. **7** 870
[10] Pogna E A A, Rodrı´guez-Tinoco C, Krisch M, Rodrı´guez-Viejo J, and Scopigno T 2013 *Sci. Rep.* | 3 : 2518 | DOI: 10.1038/srep02518 4
[11] Sokolov A P , Roessler E, Kiskiuk A and Quitman D 1993 *Phys. Rev. Lett.* **71** 2062
[12] Honglu Yand Ching C B 2004 *Chirality* **16** 541
[13] Vueba M L, Pina M E, Veiga F, Sousa J J and Batista de Carvalho L A E 2006 *Int. J. Pharm.* **307** 56
[14] Haida O, Suga H and Seki S 1977 *J. Chem. Thermodynamics* **9** 1133
[15] Talo´n C, Ramos M A, and Vieira S 2002 *Phys. Rev.* B **66** 012201

IOP Conf. Series: Materials Science and Engineering **54** (2014) 012002 doi:10.1088/1757-899X/54/1/012002

Determination of elastic stiffness coefficients of lead zirconate single crystals in the cubic phase by Brillouin light scattering

J H Ko[1], K Roleder[2] and A Bussmann-Holder[3]

[1]Department of Physics, Hallym University, 39 Hallymdaehakgil, Chuncheon, Gangwondo 200-702, Korea
[2]Institute of Physics, University of Silesia, ulica Uniwersytecka 4, PL-40-007 Katowice, Poland
[3]Max-Planck-Institut für Festkörperforschung, Heisenbergstr. 1, D-70569 Stuttgart, Germany

hwangko@hallym.ac.kr

Abstract. The temperature dependence of the three independent elastic constants of antiferroelectric lead zirconate single crystals was determined in the cubic, paraelectric phase by Brillouin light scattering spectroscopy. Two longitudinal elastic moduli of C_{11} and $(C_{11}+C_{12}+2\ C_{44})/2$ showed softening upon cooling toward the phase transition temperature, indicating the coupling of the acoustic waves to the polarization fluctuations of the precursor polar clusters. Among the two transverse acoustic modes, C_{44} was almost constant while $(C_{11}-C_{12})/2$ showed a noticeable softening in the paraelectric phase. This was attributed to the acoustic instability of lead zirconate toward the orthorhombic ground state.

1. Introduction

Lead zirconate ($PbZrO_3$, abbreviated as PZO) is one of the most representative antiferroelectric oxide materials intensively studied during the past few decades from both theoretical and experimental points of view [1]. It comprises one of the end members of famous piezoelectric $PbZr_{1-x}Ti_xO_3$ (PZT) complex perovskites, which exhibit excellent electromechanical properties at compositions near the morphotropic phase boundary. PZO undergoes an antiferroelectric phase transition at ~235°C (T_c) [2-3], below which a transient intermediate phase usually appears in a certain temperature range [4-12]. The existence of the intermediate phase is known to be sensitive to defects or stoichiometrical deviations [10]. In spite of extensive studies on this material, the exact nature of the phase transitions of PZO remains completely unsettled.

The antiferroelectric phase transition of PZO is accompanied by antiparallel shifts of the Pb atoms in the (110) plane, which coexist with the oxygen displacements undergoing antiparallel shifts within the (001) plane and along the *c* axis [9,13]. Lattice dynamic calculations also point out the coexistence of Γ point and *R* point instabilities [14]. In this respect, lattice dynamical studies are very important for better understanding of the nature of the antiferroelectric phase transition of PZO. Raman and infrared reflectivity measurements were carried out on PZO, which confirmed the existence of

[1] To whom any correspondence should be addressed.

precursor polar regions with broken symmetry in the paraelectric phase of PZO [15-16]. The measurement of the elastic stiffness coefficients (or elastic constants) is very important in relation to the interatomic potentials and their changes during the phase transition. Several theoretical calculations for elastic constants have been carried out, from which possible elastic constant values of PZO were suggested [14,17]. Recent Brillouin scattering studies showed that the predicted values from previous calculation studies are not consistent with the experimental results [18-19], demonstrating that the effective Hamiltonians used in these former calculations should be improved. In contrast, lattice dynamic calculations successfully reproduced the temperature dependence of part of the elastic constants. These clearly demonstrated that unusual lattice dynamics and large anharmonicity play an important role in the antiferroelectric phase transition of PZO [20].

In this paper, we report the details of the Brillouin scattering experiment for measuring the three independent elastic constants of PZO in its cubic phase. The detailed experimental method and results on the temperature dependence of the elastic constants will be reported. This study is in part complementary to the previous reports which include elastic data of PZO [18-20].

2. Experimental conditions

PZO single crystals were grown by using the flux growth method, i.e., from high-temperature solutions by means of spontaneous crystallization. The details of the crystal growth can be found elsewhere [18]. Several crystal plates were cut and polished to optical quality. The three orthogonal surfaces of the crystal plane were aligned along the (pseudo)cubic <100> directions. A conventional tandem 6-pass Fabry-Perot interferometer (TFP-1, JRS Co.) was used for the measurement of the Brillouin spectrum. The free spectral range was 15 GHz with a mirror spacing of 10 mm. The finesse of the interferometer was ~75. A diode-pumped solid state laser (Excelsior 532-300, SpectraPhysics) at a wavelength of 532 nm and a power of ~150 mW was used as an excitation light source. The details of the experimental setup have been described in Ref. 21.

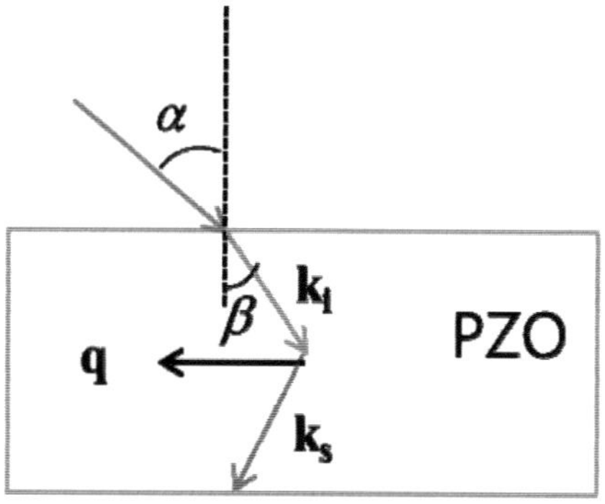

Figure 1. Forward symmetric scattering geometry.

The temperature dependence of the refractive index of PZO single crystals is usually necessary for the calculation of the elastic constants when Brillouin spectroscopy is adopted. However, a special scattering geometry, so called forward symmetric scattering geometry, may be used to obtain the elastic constants without knowing the refractive index [22]. Figure 1 shows a schematic diagram representing the forward symmetric scattering geometry. The (100) crystal plate was inserted between two cover glasses and pressed slightly by a clamp to fix it in the vertical position. Therefore, the sample is in the slightly-clamped condition rather than in the free condition. The laser beam is incident on the (100) plane with an incident angle of α (58.7±0.5°) and is refracted into the crystal with a refraction angle of β. $\mathbf{k}_i$, $\mathbf{k}_s$ and $\mathbf{q}$ denoting the wavevectors of the incident light, scattered light and acoustic phonon, respectively. Thanks to the Snell's law, the magnitude of the phonon wavevector is given by $4\pi \sin\alpha / \lambda$. By rotating the crystal plate in the vertical plane, the direction of the phonon wavevector **q** can be aligned along either <100> or <110> crystallographic axis in the cubic coordinates. Table 1 shows the phonon wavevector directions and elastic constants obtainable from each phonon propagation direction [23]. LA and TA modes denote the longitudinal acoustic and transverse acoustic modes, respectively.

A compact temperature controller (FTIR600, Linkam) was used for controlling the sample temperature. Temperature stability was good within ±0.1 K, but the temperature accuracy could not be assured due to the large aperture angle of the sample stage of FTIR600, which induced a substantial temperature gradient in the sample chamber. Therefore, the temperature dependence of elastic properties is shown in the relative temperature unit of T-T_c instead of the absolute temperature. The theoretically calculated density of 7.9826 g/cm^3 was used for the evaluation of the elastic stiffness coefficient based on the reported lattice constant of 4.1614 Å at 520 K [24].

Table 1. Formatting sections, subsections and subsubsections.

Wavevector	LA mode	TA mode
q || <100>	C_{11}	C_{44} (doubly degenerate)
q || <110>	$(C_{11}+ C_{12}+2\ C_{44})/2$	$(C_{11}- C_{12})/2$ and/or C_{44}

3. Results and discussion

Figure 2 shows the two Brillouin spectra measured at 573 K and at two scattering geometries with different phonon propagation directions. Two clear Brillouin doublets can be seen from both scattering geometries, the high-frequency mode corresponding to the LA mode and the low-frequency one corresponding to the TA mode. The elastic constants of the LA mode appropriate to each scattering geometry can be identified without any ambiguity based on Table 1. The LA mode propagating along the <100> and <110> direction refers to C_{11} and $(C_{11}+ C_{12}+2\ C_{44})/2$, respectively. The elastic constant of the TA mode at the scattering geometry of **q** || <100> is the doubly degenerate C_{44}. On the other hand, the TA mode at the scattering geometry of **q** || <110> may be either $(C_{11}- C_{12})/2$ or C_{44}. Since the temperature dependence of the TA mode at this scattering geometry is substantially different from that at the former case, we can safely identify the elastic constant of the TA mode measured at the geometry of **q** || <110> as $(C_{11}- C_{12})/2$.

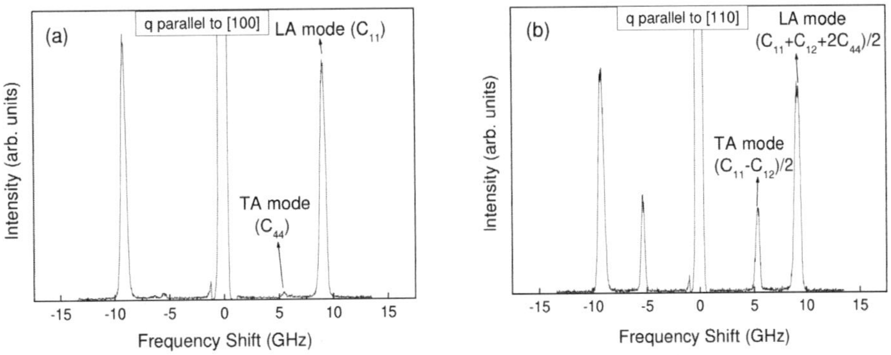

Figure 2. Brillouin spectra of PZO single crystals obtained at two different scattering geometries.

Figure 3 shows the temperature dependence of all the measured elastic constants obtained from the two scattering geometries [19]. The large experimental error bars were mostly due to the uncertainty in determining the small scattering angle (about 59°) we used in the present experiment. Table 2 shows the elastic constant values at two temperatures. The estimated error bar is approximately ±3 GPa and ±1 GPa for the LA and TA mode elastic constants, respectively.

Two longitudinal moduli, i.e., C_{11} and $(C_{11}+ C_{12}+2 C_{44})/2$ exhibit similar elastic softening upon cooling toward T_c, but the softening of C_{11} is more substantial. C_{11} decreases from ~212 GPa at (T_c+151) K to ~180 GPa near T_c. The values near T_c are still larger than the calculated values by first-principles studies (~151–161 GPa) in the cubic phase [14,17], suggesting that a more refined approach is necessary. The softening of the LA mode in the paraelectric phase was also confirmed from other perovskite polar oxides such as barium titanate ($BaTiO_3$) [25]. The softening of the LA mode was accompanied by the growth of the hypersonic damping as well as the appearance of a quasielastic central peaks [18,25]. These anomalous behaviors were attributed to the formation of precursor polar regions with locally broken symmetry and their electrostrictive interactions with the strains induced by acoustic waves in the paraelectric phase [26].

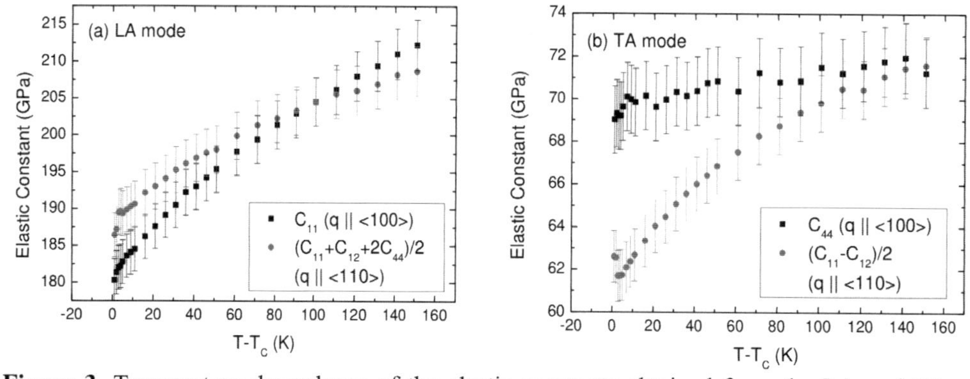

Figure 3. Temperature dependence of the elastic constants obtained from the LA and TA mode behaviors. (See Ref. [19])

The two TA modes shown in Fig. 3(b) exhibit distinct temperature dependences. C_{44} is almost constant while $(C_{11}\text{-}C_{12})/2$ shows a large softening in the paraelectric phase. C_{44} decreases from ~72 GPa at (T_c+151) K to ~63 GPa near T_c. This is in contrast to perovskite relaxors such as $Pb(Mg_{1/3}Nb_{2/3})O_3$ and $Pb(Mg_{1/3}Ta_{2/3})O_3$ [27], which exhibit noticeable softening of C_{44} upon cooling in the paraelectric phase. There are three symmetrized elastic constants in the cubic phase, that is, $C_{11}+2C_{12}$, $(C_{11}\text{-}C_{12})/2$ and C_{44} [28]. These three constants are related to the acoustic instabilities of the hydrostatic, tetragonal or orthorhombic, and rhombohedral deformation in sequence. The softening of C_{44}'s confirmed from relaxor ferroelectrics [27] were discussed in connection with the growth of the rhombohedral polar nano-regions in the paraelectric, nonpolar phase [29]. In contrast, the ground state of PZO is the orthorhombic phase. Therefore, it is reasonable that $(C_{11}\text{-}C_{12})/2$, instead of C_{44}, exhibits relatively large softening in the paraelectric phase upon cooling toward the orthorhombic phase. However, complete softening is not expected from the relevant elastic constants obtained by Brillouin light scattering because it cannot probe the zone boundary soft acoustic modes [20]. Recent lattice dynamical calculations within the polarizability model [30] showed that the temperature dependence of $(C_{11}\text{-}C_{12})/2$ could be well reproduced in the long-wavelength limit. Strong tendency towards an incipient polar instability was suggested from the calculated lattice dynamical properties. Another recent study based on inelastic X-ray scattering combined by Brillouin scattering suggested the Γ point ferroelectric instability is the single driving force for the antiferroelectric phase transition [19]. All these recent studies emphasize the important role of the anharmonic optic-acoustic mode coupling in the phase transition of PZO. More accurate determination of the elastic constants and the quasielastic central peaks in both paraelectric and antiferroelectric phases of PZO is under progress in order to find out the correlation between the acoustic instabilities and the antiferroelectric phase transition of PZO.

Table 2. Elastic constants of PZO at two temperatures

Temperature	~(T_c+151)	~T_c
C_{11} (GPa)	212.3	180.3
($C_{11}-C_{12}$)/2 (GPa)	71.6	69.0
C_{44} (GPa)	71.3	62.6

4. Conclusion

The temperature dependence of elastic constants of cubic PZO single crystals was determined by Brillouin light scattering spectroscopy. C_{11} and ($C_{11}-C_{12}$)/2 exhibited noticeable softening upon cooling toward the phase transition temperature. It suggests, in combination with the growing central peak [18], the existence of precursor polar regions with broken symmetry and their interactions with the acoustic waves. The softening of ($C_{11}-C_{12}$)/2, instead of C_{44}, was attributed to the acoustic instability related to the orthorhombic ground state. The first measurement of the elastic constants of high-quality PZO single crystals will stimulate experimental and theoretical efforts on the understanding of the nature of the antiferroelectric phase transition of this interesting material and other related perovskite polar oxides.

Acknowledgments

This research was supported by Basic Science Research Program through the National Research Foundation of Korea (NRF) funded by the Ministry of Education, Science and Technology (2013R1A1A2006582) and by the National Center for Science (NCN) in Poland within the project 1955/B/H03/2011/40.

References

[1] Liu H and Dkhil B 2011 *Z. Kristallogr.* **226** 163
[2] Roberts S 1951 *Phys. Rev.* **83** 1078
[3] Ujma Z and Hańderek J 1975 *Phys. Status Solidi a* **28** 489
[4] Jona F, Shirane G, Mazzi F and Pepinsky R 1957 *Phys. Rev.* **105** 849
[5] Tennery V 1965 J *J. Electrochem. Soc.* **112** 1117
[6] Goulpeau L 1967 *Sov. Phys. Solid State* **8** 1970
[7] Scott B A and Burns B 1972 *J. Am. Ceram. Soc.* **55** 331
[8] Fesenko O E and Smotrakov V G 1976 *Ferroelectrics* **12** 211
[9] Whatmore R W and Glazer A M 1979 *J. Phys. C* **12** 1505
[10] Roleder K and Dec J 1989 *J. Phys.: Condens. Matter* **1** 1503
[11] Viehland D 1995 *Phys. Rev. B* **52** 778
[12] Belov A A, Jeong Y H and Kang K Y 1998 *J. Korean Phys. Soc.* **32** S299
[13] Corker D L, Glazer A M, Dec J, Roleder K and Whatmore R W 1997 *Acta Cryst. B* **53** 135
[14] Waghmare U V and Rabe K M 1997 *Ferroelectrics* **194** 135
[15] Kugel G, Jankowska-Sumara I, Roleder K and Dec J 1998 *J. Korean Phys. Soc.* **32** S581
[16] Ostapchuk T, Petzelt J, Zelezny V, Kamba S, Bovtun V, Porokhonskyy V, Pashkin A, Kuzel P,

Glinchuk M D, Bykov I P, Gorshunov B and Dressel M 2001 *J. Phys.: Condens. Matter* **13** 2677
[17] King-Smith R D and Vanderbilt D 1994 *Phys. Rev. B* **49** 5828
[18] Ko J-H, Górny M, Majchrowski A, Roleder K and Bussmann-Holder A 2013 *Phys. Rev. B* **87** 184110
[19] Tagantsev A K, Vaideeswaran K, Vakhrushev S B, Filimonov A V, Burkovsky R G, Shaganov A, Andronikova D, Rudskoy A I, Baron A Q R, Uchiyama H, Chernyshov D, Bosak A, Ujma Z, Roleder K, Majchrowski A, Ko J H and Setter N 2013 *Nature Commun.* **4** 2229
[20] Bussman-Holder A, Ko J H, Majchrowski A, Górny M and Roleder K 2013 *J. Phys.: Condens. Matter* **25** 212202
[21] Kim J H, Choi J Y, Jeong M S, Ko J H, Ahart M, Ko Y H, Kim K J 2012 *J. Korean Phys. Soc.* **60** 1419
[22] Ko J H, Jeong M S, Lee B W, Kim J H, Ko Y H, Kim K J, Kim T H, Kojima S and Ahart M 2013 *Korean J. Opt. Photon.* (in press)
[23] Vacher R and Boyer L 1972 *Phys. Rev. B* **6** 639
[24] Aoyagi S, Kuroiwa Y, Sawada A, Tanaka H, Harada J, Nishibori E, Takata M and Sakata M 2002 *J. Phys. Soc. Jpn.* **71** 2353
[25] Ko J H, Kim T H, Roleder K, Rytz D and Kojima S 2011 *Phys. Rev. B* **84** 094123
[26] Rehwald W 1973 *Adv. Phys.* **22** 721
[27] Ko J H, Kim D H and Kojima S 2008 *Phys. Rev. B* **77** 104110
[28] Cowley R A 1976 *Phys. Rev. B* **13** 4877
[29] de Mathan N, Husson E, Calvarin G, Gavarri J R, Hewat A W and Morell A 1991 *J. Phys.: Condens. Matter* **3** 8159
[30] Bussmann-Holder A and Büttner H 1992 *Nature* **360** 541

Low temperature and high pressure crystals of room temperature ionic liquid: *N*, *N*-diethyl-*N*-methyl-*N*-(2-methoxyethyl) ammonium tetrafluoroborate

Hiroshi Abe[1], Yusuke Imai[1], Takahiro Takekiyo[2], Yukihiro Yoshimura[2] and Nozomu Hamaya[3]
[1]Department of Materials Science and Engineering, National Defense Academy, Hashirimizu 1-10-20, Yokosuka, 239-8686, Japan
[2]Department of Applied Chemistry, National Defense Academy, Hashirimizu 1-10-20, Yokosuka 239-8686, Japan
[3]Graduate School of Humanities and Sciences, Ochanomizu University, Tokyo 112-8610, Japan
E-mail: ab@nda.ac.jp

Abstract. Crystals of room temperature ionic liquid (RTIL) are obtained separately at low temperature or under high pressure. The RTIL is *N*, *N*-diethyl-*N*-methyl-*N*-(2-methoxyethyl) ammonium tetrafluoroborate, [DEME][BF_4]. At ambient pressure, low-temperature (LT) crystals appeared on slow cooling. By simultaneous X-ray diffraction and differential scanning calorimetry (DSC) measurements, metastable monoclinic and stable orthorhombic phases coexist in pure [DEME][BF_4]. Furthermore, the DSC thermal trace indicates that the metastable monoclinic phase was stabilized by adding water. In contrast, on compression process up to 7.6 GPa, crystallization is completely suppressed even upon slow compression. Direct observations using optical microscopy also support no crystal domain growth on compression process. High-pressure (HP) crystals at room temperature were seen only on decompression process, where two different kinds of crystals appeared subsequently. By crystal structure analysis, the LT crystal structures have no relation with the HP ones. Moreover, both metastable monoclinic phase at low temperature and higher pressure crystal has a folding molecular conformation and anti-parallel pairing of the [DEME] cation as the instability factors.

1. Introduction

Room temperature ionic liquids (RTILs) are highlighted in an electric double layer capacitor [1] and a CO_2 separation technique [2], and so on. On the other hand, nano-heterogeneity even in the liquid is a big topic on fundamental science [3]. In order to clarify the mysterious property of the RTILs, on molecular level, it is indispensable to investigate charge network and molecular orientational order based on the interaction between a cation and an anion. To remove the ambiguity in the liquid, a crystal structure of the RTILs provides a big insight to extract the molecular interactions from the molecular orientational order in crystal.

N, *N*- diethyl- *N*- methyl- *N*- (2-methoxyethyl) ammonium tetrafluoroborate, [DEME][BF_4], is one of quaternary ammonium RTILs. Firstly, for an electrochemical device possessing a wide electrochemical windows, [DEME][BF_4] was synthesized [4]. Tsuzuki *et al.* imply that oxygen of the [DEME] cation is electric negative part by density functional theory (DFT) [5]. It should be noticed that total charge of cation is plus one. Moreover, DFT calculation suggests that the [DEME] cation has a lot of degrees of freedom of molecular conformational stabilities [6]. More interestingly, the hydrophilic [DEME][BF_4] behaves more complicatedly by the hydrogen bonding of water. The complicated phase diagram in the [DEME][BF_4]-water system was determined by simultaneous X-ray diffraction and differential scanning calorimetry (DSC) measurements [7]. The attractive/repulsive interactions between them drastically change depending extensively on the water concentrations. At low temperature and ambient pressure, the crystal structures of pure [DEME][BF_4] [8] and its mixtures [9] were determined by X-ray diffraction. The crystal structures of pure [DEME][BF_4] and its mixtures were found to be monoclinic and orthorhombic phases. A significant point is that monoclinic *a*-axis corresponds to the diagonal length on *a*-*b* plane of orthorhombic lattice without the lattice mismatch.

Under high pressure and at room temperature, superpressurized phenomenon was observed by high-pressure X-ray diffraction [10, 11]. On compression process up to 7.6 GPa, there was no sharp Bragg reflections at all in [DEME][BF_4]. It's very curious that crystallization is completely suppressed in the circumstance, where Coulomb force is much dominant in the shorter molecular distance under high pressure. On a contrary, on decompression process, subsequent crystallizations were detected both by *in-situ* optical microscopic observations and X-ray diffraction. The crystal structure at 0.8 GPa is monoclinic (Z=4), while that at 0.2 GPa is monoclinic (Z=2). Here, Z reveals the number of molecules per unit cell.

In this study, further crystal structure analysis and simulations of pure [DEME][BF_4] are carried out using *ab-initio* crystal determination method in order to distinguish the high-pressure (HP) crystals from the low-temperature (LT) ones. Molecular conformations and orientational order on the lattice sites of the HP and LT crystals reveal the intrinsic lattice stability/instability.

2. Experiments

As a sample of RTIL, hydrophilic [DEME][BF_4] (Kanto Chemical Co.) is selected. For H_2O mixtures, we used distilled water (Kanto Chemical Co.).

To determine LT crystal structures and investigate phase stabilities, *in-situ* observations were performed using a simultaneous X-ray diffraction and DSC (RINT-Ultima III, Rigaku Co.) [7], where the DSC is attached on a vertical goniometer. For in-situ observations of the liquids, the sample stage is fixed horizontally, since a sealed X-ray tube and a scintillation counter move simultaneously. Cu K α radiation (λ =0.1542 nm) was used for the simultaneous measurements. Temperature range in the simultaneous measurements was from 20 to –80 °C. The cooling and heating rate were 8 °C/min and 3 °C/min, respectively. During the simultaneous measurements, dry N_2 gas was flowing at 20 cc/min in order to reduce moisture.

High-pressure X-ray diffraction measurements were carried out with the diamond anvil cell (DAC) on beamline BL-18C of the Photon Factory at the High Energy Accelerator Research Organization in Japan, where an imaging plate system (BAS2000, Fuji-Film Co., Japan) was selected to obtain two-dimensional Debye rings [12]. The incident X-ray beam was collimated to be 100 μm in diameter. The incident wavelength was estimated to be 0.061872 nm calibrated with a standard CeO_2 polycrystalline. The pressure was monitored from the spectral shift of the R_1 fluorescence line of the ruby balls inside the DAC. The observed X-ray diffraction patterns both at low temperature and under high pressure were analysed by *FOX*, which is characterized by *ab-initio* crystal structure determinations [13].

3. Results and discussion

3.1. Low-temperature crystals on slow cooling

figure 1(a) shows the thermal traces both on cooling and heating of pure [DEME][BF_4]. Obviously, crystallization occurred at around -30 °C accompanying with the exothermal peak on the thermograph

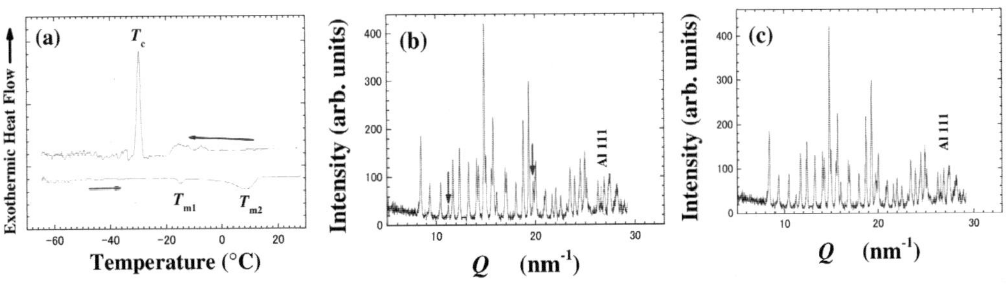

Figure 1. Thermal property of pure [DEME][BF_4] in (a) thermal traces on slow cooling and heating. Upon heating, two melting appeared. Two melting points are denoted by T_{m1} and T_{m2}, respectively. Red curves reveal the calculated X-ray diffraction patterns at -80 °C of (b) monoclinic and (c) orthorhombic phases. The arrows in the figure indicate the Bragg peaks, which are not assigned by the calculations.

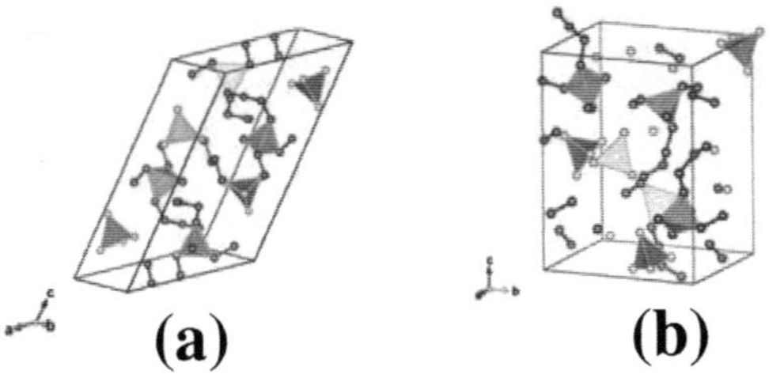

Figure 2. Low temperature crystals of (a) monoclinic and (b) orthorhombic phases in pure [DEME][BF_4].

and appearance of sharp Bragg reflections on X-ray diffraction. On the other hand, upon heating, we found two kinds of endothermic peaks in melting. Lower melting point is defined as T_{m1} and higher one is T_{m2}. As seen in crystal polymorph of [bmim][Cl] system [14, 15], monoclinic crystal is found to be metastable phase (T_{m1}), while orthorhombic one (T_{m2}) is much stabilized. The red curves in figures. 1(b) and 1(c) reveal the calculated diffraction patterns, while the black curves are the observed diffraction ones at -80 °C. By crystal structure analysis using *FOX*, two phases are determined to be $P2_1/c$ (Monoclinic, Z=4) and $P2_12_12_1$ (orthorhombic, Z=4), respectively. The detailed crystallographic data are listed in Table 1. The arrows in figure 1(b) indicate the non-assigned peaks in the monoclinic phase. The Al sample container provides the additional Bragg reflection described by Al 111 in the figures. At least, two melting points and possible two crystal structures suggest that phase coexistence is realized at the lowest temperature (-80 °C). For a further understanding, we carried out the *ab-initio* simulations, whose process is described by random molecular arrangement satisfying with the space group. Two LT crystal structures of pure [DEME][BF_4] are visualized in figures 2(a) and 2(b). Four carbons in the centre of nitrogen of a quaternary cation are described by the tetrahedron in the same manner with the green tetrahedral [BF_4] anion. Obviously, molecular orientational order of the metastable monoclinic phase is quite different from that of the stable orthorhombic one. More importantly, we found that the thermally metastable monoclinic phase can coexist without the lattice

Table 1. Crystallographic data of the low-temperature phases at ambient pressure. The weighted reliability and conventional reliability factors are expressed as wR and R, respectively.

T (K)	Space group	a (nm)	b (nm)	c (nm)	β (°)	Z	ρ (g/cm^3)	wR (%)	R (%)
200	$P2_1/c$	1.6782	0.8926	1.0130	127.0	4	1.278	20.39	16.65
200	$P2_12_12_1$	1.0128	1.3397	0.8926	90	4	1.278	19.18	15.42

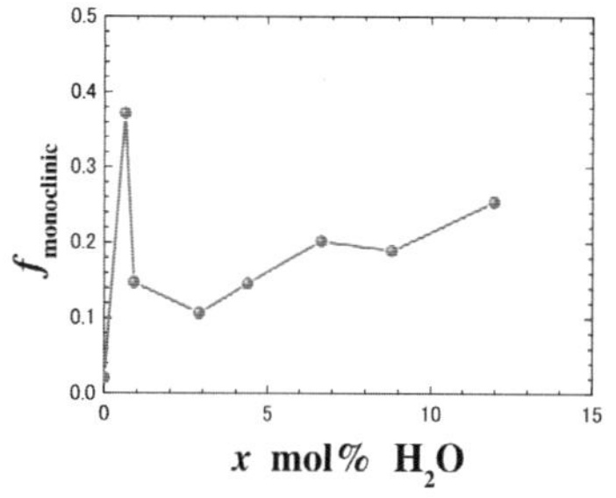

Figure 3. Monoclinic fraction, f^{monoc}, as a function of water concentration, x. Metastable monoclinic crystal is stabilized proportionally to the water concentration.

mismatch; monoclinic β angle is explained by the geometric relation as $a_{\text{M}} = \sqrt{a_{\text{O}}^2 + b_{\text{O}}^2}$, where a_{M}, a_{O} and b_{O} are lattice parameters of monoclinic a, orthorhombic a and orthorhombic b parameters, respectively.

Apart from the pure system, [DEME][BF_4]-water system was examined in order to clarify the hydrogen bonding nature of water inside the hydrophilic RTIL [4]. On the phase diagram, T_{m1} is almost constant in the water poor region, while T_{m2} decreases with increasing the water concentration. Thermodynamically, the stable orthorhombic phase in pure system became unstable by hydrogen of water. An interesting point is that metastable monoclinic phase is stabilized by additional water. For instance, a fraction of the metastable monoclinic phase, f_{monoc}, as a function of water concentration, x, is shown in figure 3. The f_{monoc} is provided by,

$$f^{\text{monoc}} = \frac{I^{\text{monoc}}}{I^{\text{monoc}} + I^{\text{ortho}}}, \tag{1}$$

where I^{monoc} and I^{ortho} reveal the decomposed peak of monoclinic and orthorhombic phases from the DSC thermal trace in melting, respectively. Certainly, endothermic peak of the orthorhombic becomes larger proportionally to the water concentration by the simultaneous measurements. It is noteworthy that extraordinal increment of f^{monoc} at around 1 mol% is caused by the specific superstructures of the [DEME][BF_4] at the water poor region [9]. The sublattice constant coincides with the average distance between water molecules at 1 mol%. Well arrayed water molecules over the medium-range could stabilize the monoclinic lattice so much. Another feature is that the lattice constant is almost the same as pure one in spite of water additive up to 20 mol% [9]. Water molecules might be intercalated into the [DEME][BF_4] crystal without the lattice expansion.

3.2. High-pressure crystals on decompression process

Upon compression process at room temperature, no crystallization occurred in pure [DEME][BF_4] and [C_2mim][BF_4], where the maximum pressure was around 8 GPa. On the other hand, on

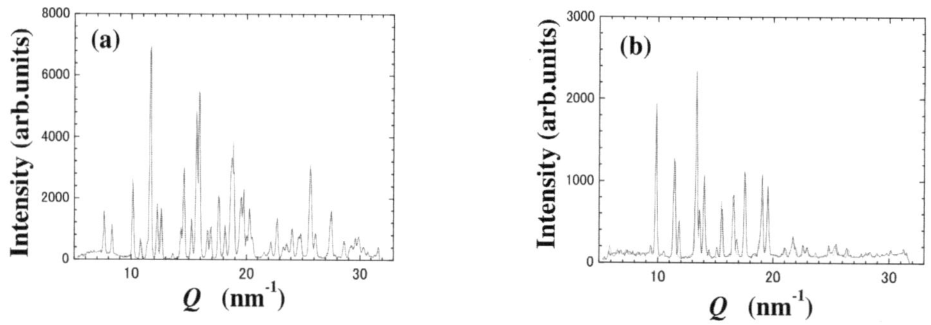

Figure 4. X-ray diffraction patterns of pure [DEME][BF_4] at (a) 0.8 GPa and (b) 0.2 GPa on decompression process at room temperature.

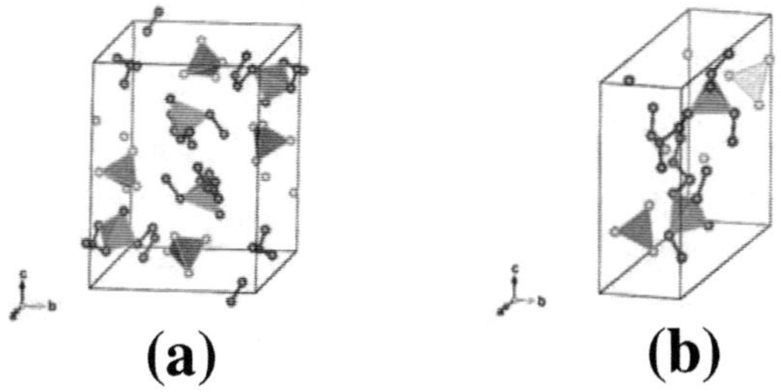

Figure 5. High pressure crystals of pure [DEME][BF_4] at (a) 0.8 GPa and (b) 0.2 GPa.

decompression process, subsequent crystallizations were observed in the RTILs [10, 11]. The non-equilibrium freezing upon compression might be relating to the conformational-driven free energy landscape [16]. But, the superpressurized effect still remains unclear. figures 4(a) and 4(b) reveal X-ray diffraction patterns on decompression process at room temperature. Judging from sharp Bragg reflections on X-ray diffraction pattern, crystallization occurred below 1.3 GPa. Further decompression causes the solid-solid phase transition. The peak positions and intensities of the Bragg reflections are quite different each other. Therefore, it was found that two different kinds of crystals appeared subsequently by reducing pressure in pure [DEME][BF_4] system. The solid-solid phase transition was also observed in the optical microscope. At the phase transition pressure, crystal domain walls were moving distinctly. Here, the crystallographic data under high pressure are listed in Table 2. Based on the HP crystal data, the simulations of the HP phases were carried out in order to obtain the molecular orientations. The molecular arrangements and orientational order at 0.8 and 0.2 GPa are visualized as shown in figures 5(a) and 5(b), respectively. By losing *c*-glide, phase I (0.8 GPa) transforms to phase II (0.2 GPa) without varying the β angle. Consequently, *b* lattice parameter of phase II (Z=2) becomes the half value of phase I (Z=4).

Table 2. Crystallographic data of the high-pressure phases at room temperature.

P (GPa)	Space group	a (nm)	b (nm)	c (nm)	β (°)	Z	ρ (g/cm^3)	$_wR$ (%)	R (%)
0.8	$P2_1/c$	0.791	1.081	1.342	107.9	4	1.42	22.3	18.6
0.2	$P2_1$	1.032	0.521	1.147	107.6	2	1.32	27.9	23.9

3.3. Comparison between low-temperature and high-pressure crystals

In order to clarify thermal and pressure effects associating with the molecular interactions, we compare the crystal structures both of LT and HP phases. Two HP phases have no relation with the LT ones. Space group of the HP crystals was found to be monoclinic $P2_1/c$. Although the space group of the HP crystal (0.8 GPa) is the same as one of the LT crystals, lattice constants and β angle are entirely different as listed Tables 1 and 2. Therefore, we deduce that molecular-distance tunable pressure plays a different role, compared with the thermal effect. As another approach, degrees of freedom of cation conformations and molecular orientational order are a clue to obtain the molecular interactions at the each circumstance. From the viewpoint of molecular conformations, the cation conformations on the lattice sites are provided by the simulations. Conformations of [DEME] cation of metastable monoclinic (-80 °C), orthorhombic (-80 °C), monoclinic (0.8 GPa) and monoclinic (0.2 GPa) phases are shown in figures 6(a)-6(d), respectively. The peculiar feature is that cation foldings are demonstrated in metastable LT crystal (figure 6(a)) and highly packed HP one at 0.8 GPa (figure 6(c)). The characteristic folding cation is a key to understand the phase instability, since a *zigzag* sequence of cation is commonly preferred to minimize the electrostatic energy in the crystals of the conventional RTILs. In fact, the stable orthorhombic of the LT phase has no foldingconformer of cation (figure 6(b)).

In the charge-network preferred environment, a cation folding could be not formed for the well-arranged molecules in crystal. If we assume that the folding cation is regarded as the instability factor, instability might be induced in the higher pressure phase at 0.8 GPa. At 0.8 GPa, a packing efficiency has a priority against the electrostatic stabilization. In the same manner with the folding factor, molecular orientational order is a significant element considering the phase stability including a dipole-dipole interaction. For simplicity, [DEME] cation in crystal is represented only by oxygen and

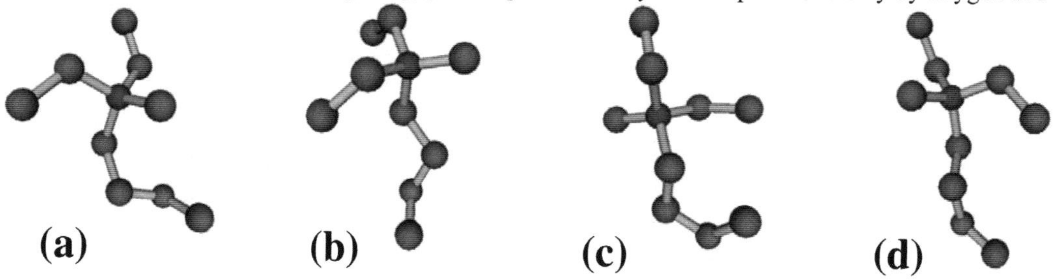

Figure 6. Conformations of [DEME] cation on the lattice sites of (a) monoclinic (-80 °C), (b) orthorhombic (-80 °C), (c) monoclinic (0.8 GPa) and (d) monoclinic (0.2 GPa) phases.

nitrogen in figures 7(a)-7(d). As theoretically predicted [5], oxygen (red circles) is electro negative part of cation, while nitrogen (silver circles) is located in the center of positive charge. Green circles are boron, which is central position of an anion. Surprisingly, anti-parallel pairing of the cation is seen clearly in the metastable LT phase (figure 7(a)). This means that charge network breaking occurs in the vicinity of the cation paring. On the other hand, a simple charge network is developed in the stable orthorhombic phase (figure 7(b)). Under high pressure, the specific cation pairing is also seen in the higher pressure phase at 0.8 GPa (figure 7(c)). It is clear that a cation-anion pair is interrupted by the cation pairing even under high pressure. At lower pressure (0.2 GPa), a cation and anion array such as a normal ionic crystal structure was obtained in figure 7(d). In addition to the folding conformation of cation, the anti-parallel pairing of cation is the second indicator of phase instability.

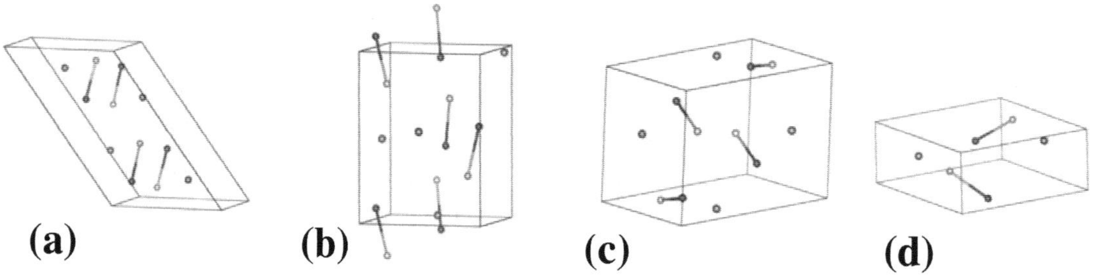

Figure 7. Cation orientational order of (a) monoclinic (-80 °C), (b) orthorhombic (-80 °C), (c) monoclinic (0.8 GPa) and (d) monoclinic (0.2 GPa) phases. Red and grey circles stand for oxygen and nitrogen, respectively. For simplicity, carbon, hydrogen and fluorine are omitted in the figures. The oxygen is electro negative part of cation, while nitrogen locates in the centre of positive charge. Green circles reveal boron of BF_4^- anion.

4. Concluding remarks

Crystal structures of LT and HP phases are distinguished by further structure analysis and simulations. Reflecting on the molecular interactions, the LT crystal structures are different from the HP ones. By the detailed structure analysis, cation folding and anti-parallel pairing of cation appeared in unstable phases both at low temperature and under high pressure (instability factors). The intrinsic interactions provide an influence on the crystallization process. Normal crystallization occurs on slow cooling, although crystallization is completely suppressed on compression process (superpressurize phenomenon). The different structures, instability factors and superpressurized phenomenon are originated from a balancing among charge network, molecular orientational order, cation folding/unfolding and packing efficiency. Since pressure provides quite different results in pure [DEME][BF_4], pressure might be regarded as a significant external field to extract the inherent property of the RTILs.

Acknowledgments
Authors appreciate Ms. Yasaka and Mr. Kishi of Rigaku Co for helpful experimental supports. By Photon Factory Research Project (Proposal #: 2010G546), we performed the experiments of high-pressure X-ray diffraction.

[1] Lockett V, Sedev R, Ralston J, Horne M and Rodopoulos T 2008 *J .Phys. Chem. C* **112** 7486
[2] Blanchard L A, Hancu D, Beckman E J and Brennecke J F 1999 *Nature* **399** 28
[3] Russina O, Triolo A, Gontrani L, Caminiti R, Xiao D, Hines Jr. L G, Bartsch R A, Quitevis E L, Plechkova N and Seddon K R 2009 *J. Phys.: Condens. Matter* **21** 424121
[4] Sato T, Masuda G and Takagi K 2004 *Electrochim. Acta* **49** 3603
[5] Tsuzuki S, Hayamizu K, Seki S, Ohno Y, Kobayashi Y and Miyashiro H 2008 *J. Phys. Chem. B* **112** 9914
[6] Takekiyo T, Imai Y, Abe H and Yoshimura Y 2012 *Adv. Phys. Chem.* 829523
[7] Abe H, Yoshimura Y, Imai Y, Goto T and Matsumoto H 2009 *J. Mol. Liq.* **150** 16
[8] Imai Y, Abe H, Goto T, Yoshimura Y, Kushiyama S and Matsumoto H 2008 *J. Phys. Chem. B* **112** 9841
[9] Imai Y, Abe H and Yoshimura Y 2009 *J. Phys. Chem.* B **113** 2013
[10] Yoshimura Y, Abe H, Imai Y, Takekiyo T and Hamaya N 2013 *J. Phys. Chem. B* **117** 3264
[11] Yoshimura Y, Abe H, Takekiyo T, Shigemi M, Hamaya N, Wada R and Kato M 2013 *J. Phys. Chem. B* **117** 12296
[12] Shimomura O, Takemura K, Fujihisa H, Fujii Y, Ohishi Y, Kikegawa T, Amemiya Y and Matsushita T 1992 *Rev. Sci. Instrum.* **63** 967
[13] Favre-Nicolin V and Černŷ R 2002 *J. Appl. Crystallogr.* **35** 734
[14] Hayashi S, Ozawa R and Hamaguchi H 2003 *Chem. Lett.* **32** 498
[15] Holbrey J D, Reichert W M, Nieuwenhuyzen M, Johnston S, Seddonbc K R and Rogers R D 2003 *Chem. Commun.* 1636
[16] Debenedetti P. G. and Stillinger F. H. 2001 *Nature* **410** 259

Electrode dependent interfacial layer variation in Metal-Oxide-Semiconductor capacitor

I-S Park[1,2], Y C Jung[1], M Lee[1], S Seong[1], and J Ahn[1]
[1] Department of Materials Science & Engineering, Hanyang University, Seoul, Korea
[2] Institute of Nano Science and Technology, Hanyang University
222 Wangsimni-ro, Seongdong-gu, Seoul 133-791, Korea

E-mail: jhahn@hanyang.ac.kr

Abstract. The interfacial layer between oxide and semiconductor in metal-oxide-semiconductor (MOS) capacitors depends on the metal electrode material. The metal/HfO_2/Si and metal/HfO_2/Ge capacitor were made using an atomic layer deposited HfO_2 dielectric films and Mo, Ru, and Pt electrodes above Si substrate and Ti, Ru, and Pt electrodes above Ge substrate. The measured saturation capacitance was varied with electrode and evaluated to capacitance equivalent thickness (CET). In Si-based MOS capacitor, the CET value of the capacitor with Pt electrode is larger than those with Mo and Ru electrode. In addition, the CET is 27.4 A, 38.2 A, and 30.8 A for Ti, Ru, and Pt electrode, respectively, for Ge-based MOS capacitors. The CET variation with electrode is attributed the variation of dielectric constant of HfO_2 dielectric and the difference of interfacial layer. The CET variation is well in agreement with the interfacial layer thickness taken by a transmission electron microscopy. The thickness variation of interfacial layer results from the oxygen gettering ability of the electrode even though they are apart.

1. Introduction

Si-based metal-oxide-semiconductor (MOS) transistors have faced a challenge for continuous and exponential scaling-down of feature size. With this rapid scaling of MOS transistors, that is, the decrease of channel length and gate dielectric thickness, new materials, processes, and structures have been introduced to the IC manufacturing industry [1]. As new materials, high-mobility channels, high dielectric constant (k) gate dielectrics, and metal gate electrodes are under research and development to replace Si channel, SiO_2 dielectric, and doped-polycrystalline Si electrode, respectively. With the successful introduction of high-k and metal gate (HKMG) beyond 45 nm technology [2, 3], high mobility channels such as Ge and GaAs have been focused as a further solution for the carrier mobility enhancement in logic devices. On the high-k/semiconductor interface, the native oxide, playing a role of additional dielectric with high-k oxide, is a layer to decrease the capacitance of MOS transistor. With scaling-down of devices, the interfacial native oxide should be controlled and reduced for higher capacitance [4, 5]. When the devices are processed with the same high-k deposition and post-annealing, the capacitance of capacitor is strongly dependent on the metal electrode materials. The value of capacitance is easily compared in term of equivalent oxide thickness (EOT) or capacitance equivalent thickness (CET). The electrode dependent CET of capacitor has also been reported [6, 7].

IOP Conf. Series: Materials Science and Engineering **54** (2014) 012004 doi:10.1088/1757-899X/54/1/012004

Atanassova *et al.* reported the EOT and dielectric constant variation of Ta_2O_5 MOS capacitors with Al, Au, W, TiN, and TiN/W electrodes, and suggested that these variations are related to the gate electrode/Ta_2O_5 interaction during electrode deposition [6]. Kim *et al.* also reported the EOT decrement of HfO_2 and ZrO_2 MOS capacitors when Ti is used as a top electrode compared to those of Al and Pt, and claimed that this decrement comes from the decomposition of the SiO_2 interfacial layer between high-k dielectric and Si substrate [7].

Atomic layer deposition (ALD) has been widely employed to making high-k gate oxides [1, 4, 5]. In ALD, the film grows through a self-limiting surface reaction between precursors that are alternately supplied into a substrate. The precursors are separated by inert gas purging to remove the physisorbed precursor and to block the gas-phase reaction between precursors. Even though the ALD method has the disadvantage of low growth rate, it offers high-quality thin films, excellent atomic-level thickness controllability and uniformity in a wide surface, high conformality, and excellent repeatability.

Metal for gate electrode applications requires a suitable work function and thermal stability [8, 9]. Single metal electrodes with a higher work function similar to conduction band such as Pt and Ru are the promising candidates for p-MOS devices and that with a lower work function similar to valence band such as Ti are for n-MOS. Mo electrode has applied for p-MOS, n-MOS and midgap devices.

In this work, the capacitance variations of both Si- and Ge-based MOS capacitors were investigated with the structures of metal/ALD-HfO2/Si and metal/ALD-HfO2/Ge using capacitance-voltage (C-V) measurements. The capacitance was dependent on a metal electrode even though the capacitors were processed indentically except for the electrode specimen. To investigate the structural status of interfacial layers concerning to this capacitance variation, cross-sectional transmission electron microscopy (TEM) analyses were also conducted.

2. Experiments

The substrates used in this study were p-type (100) Si wafers with B of 3×10^{15} atoms/cm^3 doping concentration and p-type (100) Ge wafers with Ga of 4.5×10^{17} atoms/cm^3 each other. Prior to HfO_2 deposition on Si wafers, standard RCA cleaning with final 100:1 HF-dip for 10 sec was performed to remove the native oxide and also on Ge wafers, wet chemical cleaning was performed with 100:1 HF dipping, followed by DI water rinse for five cycles [10]. High-k HfO_2 dielectric films were deposited in a travelling-wave type ALD reactor (made by CN1) using a liquid $Hf[NCH_3C_2H_5]_4$ precursor and H_2O oxidant. Hf precursor was volatilized from the canister heated at 60°C and was carried into the reaction chamber with a carrier gas of N_2. Nitrogen was also used as a purge gas. The canister containing H_2O was kept at room temperature and no carrier gas was used to deliver it to the ALD reactor due to its high vapor pressure. After HfO_2 film deposition, square-type gate pattern was defined using a conventional photolithography process. And then, top electrodes such as Mo, Ru, Pt, and Ti were deposited on HfO_2 film and photoresist by a DC magnetron sputtering system at room temperature. In case of Ti electrode, Pt was also stacked on it to prevent its surface oxidation. The base pressure of the sputter system was less than 2×10^{-6} torr and deposition was performed at 3×10^{-3} torr. The back surfaces of the substrates were metallized to provide ohmic contact to semiconductor substrate for electrical measurements. The fabricated MOS capacitors were treated by forming gas anneal (FGA) for 30 min in N_2 ambient with 5 % H_2 at 400°C.

The C-V measurements were performed in parallel mode at 1 MHz frequency using a HP4284 LCR meter by sweeping from 2 V to -2 V. For the cross-sectional analyses for the interfacial layer (IL), TEM measurements were performed for all MOS capacitors.

3. Results and Discussion

Figure 1 shows the typical C-V curves of metal/HfO_2/Si MOS capacitors with Mo, Ru, and Pt metal electrodes. The physical thickness of high-k HfO_2 on native oxide is the same 5 nm. The saturated capacitance at -2 V could be evaluated into CET for comparison and the measured CET is 19.5 A, 20.7 A, and 26.0 A for Mo, Ru, and Pt electrode, respectively. It is observed that the CET value of the capacitor with Pt electrode is larger than those with Mo and Ru electrode, even though all capacitors were processed indentically except for the material specimen of metal electrode. Another apparent feature is the shift of C-V curves with metal electrode material. The C-V curve of capacitor with Pt electrode is much shifted positive direction with respect to those with Mo and Ru electrode. The shift of C-V curves is relatively in agreement with the work function variation of the metal materials [4].

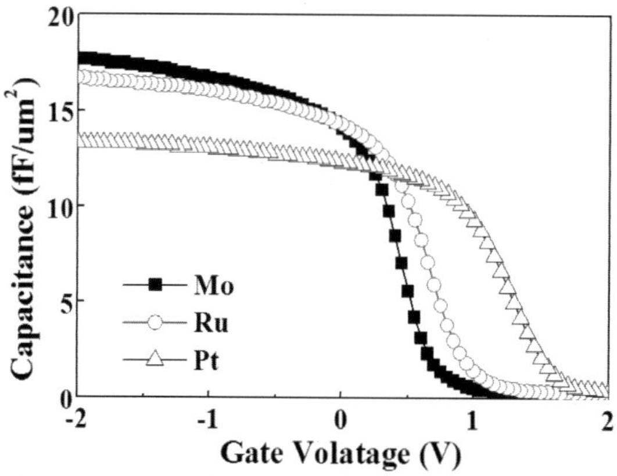

Figure 1. Electrode material dependent C-V characteristics of metal/HfO_2/Si MOS capacitors with Mo, Ru, and Pt electrode, respectively. The high-k HfO_2 films are 5 nm thick.

In addition, Fig. 2 shows the typical C-V curves of metal/HfO_2/Ge MOS capacitors with Ti, Ru, and Pt metal electrodes. These curves of Ge-based MOS capacitors are very different with those of Si-based capacitors in Fig. 1. It is found a stretch-out in the C-V curve meaning many interfacial traps and a higher value of in the inversion region indicating an additional and parasitic capacitor generated near a Ge substrate. On the parasitic capacitance at inversion region for capacitors with Ru electrode, Cmin/Cmax for Si capacitor is 1.8 % but that for Ge is 18.2 % where Cmin is the minimum capacitance and Cmax is the maximum. These results indicate the unstable GeO_x interfacial layer formation on Ge substrate. It is also observed the saturated capacitance at -2 V is dependent on the metal electrode and the measured CET is 27.4 A, 38.2 A, and 30.8 A for Ti, Ru, and Pt electrode, respectively. Interestingly, the CET value of the capacitor with Ru electrode is larger than that of Pt electrode. It is the opposite result with that for Si substrate. For the shift of C-V curves, it is very little in case of Ge substrate due to the Fermi level pinning [11].

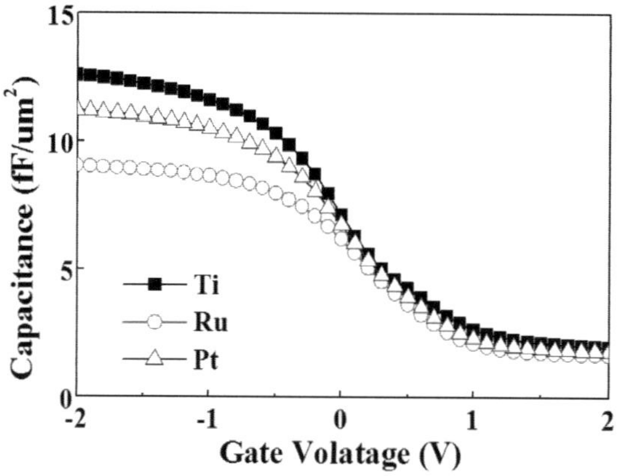

Figure 2. Electrode material dependent C-V characteristics of metal/HfO_2/Ge MOS capacitors with Ti, Ru, and Pt electrode, respectively. The high-k HfO_2 films are 5 nm thick.

IOP Conf. Series: Materials Science and Engineering **54** (2014) 012004 doi:10.1088/1757-899X/54/1/012004

As the capacitance of simple capacitor is expressed in form of C = kA/d where k is the dielectric constant of dielectric, A is the effective area of capacitor, and d is the thickness of dielectric. And, CET can be evaluated as CET = $3.9A/C_m$, where k of SiO_2 is 3.9 and C_m is the measured capacitance in maximum. When three dielectrics are stacked in parallel, the total CET = $CET_1 + CET_2 + CET_3$, where CET_n is CET of n-th dielectric. In the MOS capacitor, usually, three dielectric layers are can be supposed, that is, 1) the bottom IL between oxide and semiconductor, 2) main dielectric (it will be denoted as dielectric), and 3) the top IL between metal and oxide.

To identify ILs and dielectric affecting to capacitance variation with metal electrode, TEM images of MOS capacitors were taken as shown in Fig. 3. It is found that the Ge surfaces under bottom IL are much rougher than Si surfaces due to the softness of Ge against wet HF chemical. At first, it is observed that top ILs show rough surface compared to bottom ILs but there is not so apparent layer in all images in Fig. 3. For the HfO_2 dielectrics, their structures were not crystallized on Si and Ge substrates due to the low thermal budget. Their values of dielectric constant on Si are 17.7, 18.8, and 23.2, respectively, which were evaluated from the relationship between EOT and the physical thickness of HfO_2 films [4]. Considering the dielectric constant of HfO_2, for example 17.7, the contribution of HfO_2 film with 4.5 A (k(HfO_2)/k(SiO_2)) corresponds to 1 A of CET. The higher the dielectric constant of HfO_2 is, the lower its contribution to CET is. This result indicates that the dielectric constant of dielectric is strongly dependent on the top metal electrode. This dependence comes from the deposition process of electrode and post-annealing [6], which attribute to the inter-diffusion of elements in both dielectric and electrode.

For the bottom ILs, they show obviously different thickness in the TEM images with electrode. In case of Si substrate, the IL with Pt electrode is about 15 A whereas those with Mo and Ru electrodes are about 7-8 A. This difference is well in agreement with the difference of CET in Fig. 1. For Ge substrate, the ILs below Ti and Pt are much thinner than IL below Ru, so that CET of the capacitors with Ti and Pt result in a lower value. Ti film as well as Ba, Ti, Zr, Hf, V, Nb, and Ta is well-known as one of oxygen-getter materials [12]. Assuming the gettering ability of Ti film whether it is apart with IL and so IL becomes thinner after electrode deposition and FGA, other metal layers such as Mo, Ru, and Pt can also getter the oxygen in ILs. However, the gettering effect is different with the specimen of bottom IL. For example, Ru film can getter the IL on Si but not much on Ge as shown in Figures. 3(b) and 3(f). It is found that Mo and Ru films are efficient at gettering oxygen from SiO_2 IL and Ti and Pt films are from GeO_x IL. In addition, the oxygen gettering of electrode from the oxide film, HfO_2, gives rise to variation of dielectric constant of dielectric film.

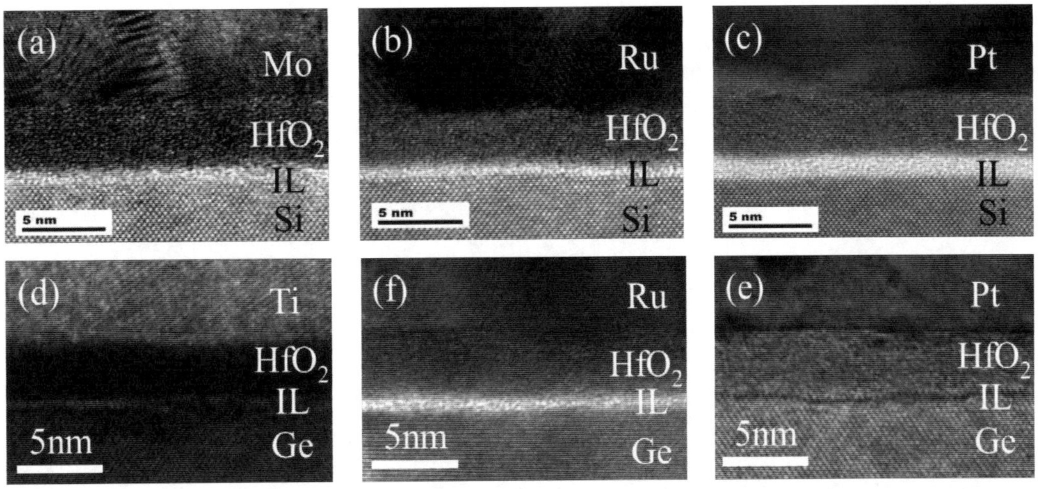

Figure 3. Cross-sectional TEM images of metal/HfO_2/Si MOS capacitors with the electrode of (a) Mo, (b) Ru, and (c) Pt and metal/HfO_2/Ge capacitors with (d) Ti, (e) Ru, and (f) Pt electrode, respectively.

4. Conclusion

The electrode dependent ILs in MOS capacitors of metal/HfO_2/Si and metal/HfO_2/Ge are investigated by using C-V measurements and cross-sectional TEM analyses. The metal/ALD-HfO_2/Si capacitors with Mo, Ru, and Pt electrode and metal/ALD-HfO_2/Ge capacitors with Ti, Ru, and Pt electrode were fabricated. The measured values of CET were varied with electrode material. In Si-based MOS capacitor, the CET value of the capacitor with Pt electrode is larger than those with Mo and Ru electrode. In addition, the CET is 27.4 A, 38.2 A, and 30.8 A for Ti, Ru, and Pt electrode, respectively, for Ge-based MOS capacitors. The CET variation is attributed to the variation of dielectric constant of HfO_2 film and the thickness difference of IL. The CET variation is well in agreement with the thickness of ILs. Even though electrode and IL are apart, the oxygen-gettering of the electrode affects to IL. The choice of electrode is a promising approach to control the state of IL on semiconductor for the higher capacitance.

Acknowledgments

This work was supported by the IT R&D program of MOTIE (KI002083, Next-Generation Substrate Technology for High Performance Semiconductor Devices) and Basic Science Research Program through the National Research Foundation of Korea(NRF) funded by the Ministry of Education(2012R1A6A1029029).

References

[1] Wilk G D, Wallace R M and Anthony J M 2001 *J. Appl. Phys.* **89** 5243
[2] Veloso A, Hoffmann T, Lauwers A, Yu H, Severi S, Augendre E, Kubicek S, Verheyen P, Collaert N, Absil P, Jurczak M and Biesemans S 2007, *Sci. Technol. Adv. Mater.* **8** 214
[3] Ishimaru K 2008 *Solid-State Electron.* **52** 1266
[4] Park I-S, Lee T, Ko H, Ahn J 2006 *J. Kor. Phys. Soc.* **49** S760
[5] Park I-S, Choi Y, Nichols W T, Ahn J 2011 *Appl. Phys. Lett.* **98** 102905
[6] Atanassova E, Paskaleva A, Novkovski N and Georgieva M 2005 *J. Appl. Phys.* **97** 094104
[7] Kim H , McIntyre P Chui C O, Saraswat K C and Stemmer S 2004 *J. Appl. Phys.* **96** 3467
[8] Yeo Y-C 2004 *Thin Solid Films* **462-463** 34
[9] Cabral C Jr., Lavoie C, Ozcan A S, Amos R S, Narayanan V, Gusev E P, Jordan-Sweet J L and Harper J M E 2004 J. Electrochem. Soc. **151** F283
[10] Deegan T and Hughes G 1998 *Appl. Surf. Sci.* **123/124** 66
[11] Chandra S V J, Fortunato E, Martins R, Choi C J 2012 *Thin Solid Films* **520** 4556
[12] Cardarelli F 2008 *Materials Handbook: A Concise Desktop Reference 2nd ed.* (London: Springer)

Tsukuba International Conference on Materials Science 2013 IOP Publishing
IOP Conf. Series: Materials Science and Engineering **54** (2014) 012005 doi:10.1088/1757-899X/54/1/012005

Crossover from diffuse to sharp phase transition by electric field in $0.955Pb(Zn_{1/3}Nb_{2/3})O_3$-$0.045PbTiO_3$

Shinya Tsukada[1], Yukikuni Akishige[1], Tae Hyun Kim[2], and Seiji Kojima[2]

[1]Faculty of Education, Shimane University, Matsue, Shimane 690-8504 , Japan
[2]Graduate School of Pure and Applied Sciences, University of Tsukuba, Tsukuba, Ibaraki 305-8573, Japan

E-mail: tsukada@edu.shimane-u.ac.jp

Abstract. Effect of an electric field along [100] direction on diffuse phase transitions in a relaxor ferroelectric $0.955Pb(Zn_{1/3}Nb_{2/3})O_3$-$0.045PbTiO_3$ (PZN-0.045PT) crystal is studied through dielectric properties and a longitudinal acoustic (LA) phonon. Compared with the diffuse phase transition observed in zero-field cooling in PZN-0.045PT, it shows two sharp phase transitions under field cooling. This result indicates that inhomogeneity is reduced by the electric field in the way that electric field grows the static polar nanoregions and microdomains into larger ferroelectric domains. From a thermodynamic point of view, the electric field enhances the relaxation from the nonequilibrium polymorphic states to the equilibrium homogeneous state. At the same time, ferroelectric tetragonal phase is stabilized with increasing an electric field along [100].

1. Introduction

A special class of ferroelectric materials, relaxor ferroelectrics, has attracted much attention due to their giant dielectric and electromechanical responses [1, 2]. The responses are as a result of polar fluctuations of intrinsic inhomogeneity. Such phenomena in inhomogeneous systems have attracted a great deal of research in the field of condensed matter physics, because intrinsic inhomogeneity in complex oxides sometimes induces some kinds of unusual behavior [3, 4]. In relaxor ferroelectrics, local polar regions of several nanometers wide, called polar nanoregions (PNRs), static nanodomain and microdomain structures are observed as intrinsic inhomogeneity; that is, relaxor ferroelectrics demonstrate structural hierarchy. Although the response of homogeneous inorganic dielectric materials to an external electric field is usually related to ionic displacements, relaxor ferroelectrics show high-order structures at a mesoscopic scale in addition to ionic displacements—so responses to an electric field are enhanced by their fluctuations. Thus, relaxor ferroelectrics are an intriguing system in which intrinsic inhomogeneity can lead to a large-magnitude response to an external electric field.

$Pb(Zn_{1/3}Nb_{2/3})O_3$ and its solid solution with typical ferroelectric $PbTiO_3$, $(1\text{-}x)Pb(Zn_{1/3}Nb_{2/3})O_3\text{-}xPbTiO_3$ (PZN-*x*PT), are one of the most extensively investigated relaxor systems, which are distinguished from normal ferroelectrics by the broad and frequency-dispersive maximum in the temperature dependence of dielectric constant [1, 2]. The crystal structure of PZN-*x*PT is characterised by inhomogeneous arrangement of Zn^{3+}, Nb^{5+}, and Ti^{4+} on crystallographically equivalent sites in the perovskite structure. The disorder due to the different ions at equivalent lattice sites generates PNRs at much higher temperatures than a ferroelectric phase-transition temperature, leading to structural hierarchy in relaxor ferroelectrics.

Various investigations of relaxor ferroelectrics have been undertaken to understand the curious phenomena. One intriguing macroscopic phenomenon is phase transitions. We have investigated the role of PNRs in the ferroelectric phase transition in relaxor ferroelectrics through dielectric and acoustic phonon measurements [5–12]. Our results indicate that the ferroelectric phase transition as well as the dynamics of PNR is sensitive to composition of $PbTiO_3$ and thermal treatments. Moreover, since Kutnjak *et al.* pointed out the enhanced polarization fluctuation around a critical end point in temperature-electric field phase diagram, studies on the electric field induced phenomena have been performed by many researchers, which indicates that relaxor ferroelectrics also very sensitive to an electric field [13–18]. As most of works are on the static properties, at the present stage, it is necessary to investigate the phase transition from a dynamical point of view. Thus, relaxor dynamics are discussed in the present study by measuring Brillouin scattering and dielectric properties under an electric field in a PZN–0.045PT single crystal. By this complementary measurements enable us to discuss the dynamics in the broad frequency range up to the GHz range.

2. Experimental

A single crystal of PZN–0.045PT was grown using the high-temperature flux technique with PbO-based fluxes (Microfine Materials Technologies). Each surface of the crystals was a (100) plane of pseudo-cubic orientation. The LA phonon close to the Γ point of reciprocal space was observed through inelastic light scattering. The light scattering was measured with a high-contrast 3+3-pass tandem Fabry–Pérot interferometer (JRS Scientific Instruments) combined with an optical microscope (Olympus). The PZN–0.045PT single crystal was placed inside a cryostat cell (Linkam Scientific Instruments) to allow the sample temperature to be adjusted between 80 K and 873 K. The measurements were performed on cooling with and without applying an external electric field by using a DC power source (Matsusada Precision). A diode-pumped solid-state laser (Coherent Inc.) with single-frequency operation at 532 nm and an output power of 100 mW was used. A 180°-scattering geometry without a polarizer was adopted for observing the LA phonon.

The complex dielectric constant ($\varepsilon^* = \varepsilon' - j\varepsilon''$) of a PZN–0.045PT single crystal was measured using an *LCR* meter (HP) in the [100] direction. The measurements were performed while cooling at a rate of 2 K min^{-1} in a homemade furnace. An external field was applied for the dielectric measurements by a DC power supply (Matsusada Precision). To avoid applying an electric field on the *LCR* meter, a homemade circuit containing condensers as low-frequency filters was set between the sample and the *LCR* meter [19].

3. Results and Discussion

3. 1 Brillouin scattering

Figure 1 (a) shows the light scattering spectra at three temperatures with and without an electric field of 500 V/cm. The spectrum consists of a longitudinal acoustic (LA) phonon at 42 GHz, which is related to the complex elastic stiffness constant $c_{11}^* = c_{11}' - jc_{11}''$, and a central peak at 0 GHz, which is related to relaxation in the GHz range. The LA phonon is apparently temperature and electric field dependent, and the dependences are shown in Figure 1 (b). Figure 1 (b) shows the contour map of light scattering spectra as functions of the frequency shift and temperature. The data on ZFC shows a clear

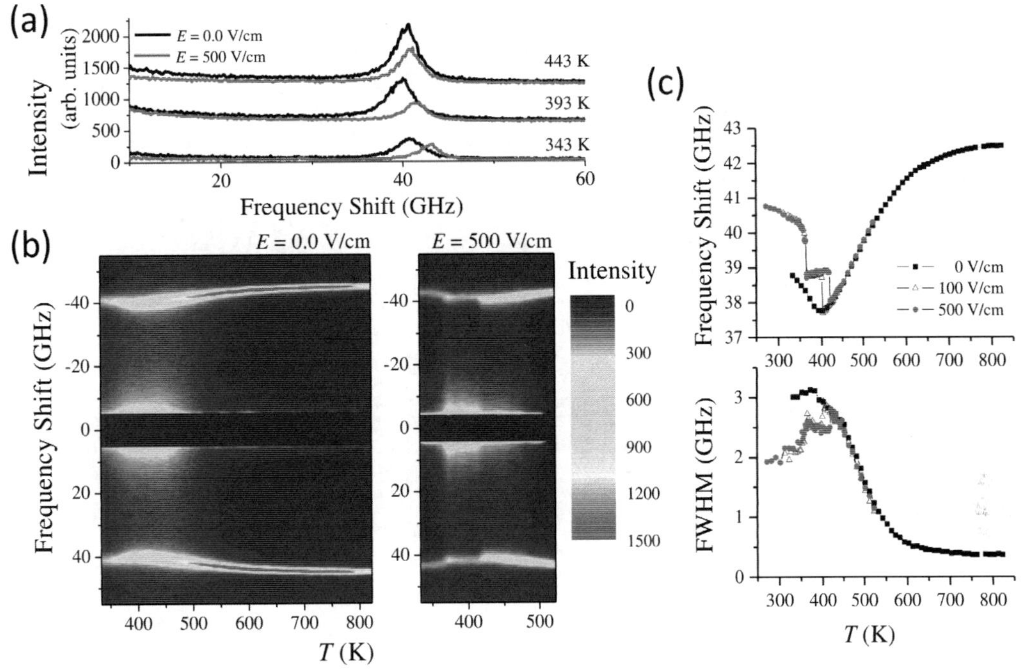

Figure 1. Inelastic light scattering in a GHz range from a PZN–0.045PT single crystal on ZFC and FC (E= 500 V/cm). (a) Inelastic light scattering spectra at three different temperatures. The peak around 42 GHz denotes the scattering from the LA phonon. Note that LA phonon propagates perpendicular to E. (b) Contour map of inelastic light scattering intensity versus temperature and frequency shift in $x(z\ y+z)\bar{x}$ scattering geometry (Free Spectral Range = 75 GHz, scan range = 70.5 GHz). (c) Frequency shift, ν_B, and FWHM, Γ, of the LA phonon under 0, 100, and 500 V/cm, which are determined by fitting the spectra.

anomaly in the LA phonon around 400 K: the LA phonon softens toward T_C. At the same time, the peak broadens towards T_C. The anomalous behavior is typical of ferroelectric phase transitions in perovskite ferroelectrics through strain–polarization coupling of free energy [20, 21]. However, the temperature range over which the phonon anomaly observed in a paraelectric phase is especially large for the cubic perovskite structure, reflecting the characteristics of relaxor ferroelectrics. The LA phonon on FC shows different phase transition behaviors: two anomalies appear at 420 K and 363 K, which correspond to the phase transitions from cubic phase to tetragonal one, then to rhombohedral. Note that the electric field at critical end point in PZN–0.045PT is around 1000 V/cm (E_{CEP}). To extract the frequency shift and the width of the LA phonon from the light scattering spectra, we used the Voigt function. The width of a Gaussian component in the Voigt function was fixed as an instrumental function. The computed parameters of the acoustic phonons, including Brillouin shift, ν_B, and full-width at half-maximum (FWHM), Γ, are shown in Figure 1 (c) as a function of temperature for the PZN–0.045PT under E= 0, 100, and 500 V/cm, respectively. The ν_B and Γ represent the characteristic tendencies of the LA phonon (i.e., softening and broadening over a wide temperature range more clearly than Figure 1 (b). Above T_C, typical perovskite ferroelectrics do not show an anomaly in the LA phonon, or only show anomalies in a narrow temperature range less than 10 K wide, because the piezoelectric coupling is prohibited in their paraelectric phases due to the symmetry restriction [21]. Consequently, the LA phonon anomaly in the relaxor ferroelectric PZN–0.045PT indicates that strong coupling between strain fluctuation and polarization fluctuation exists in the relaxor ferroelectrics [11, 22]. This characteristic behavior does not change with the application of an

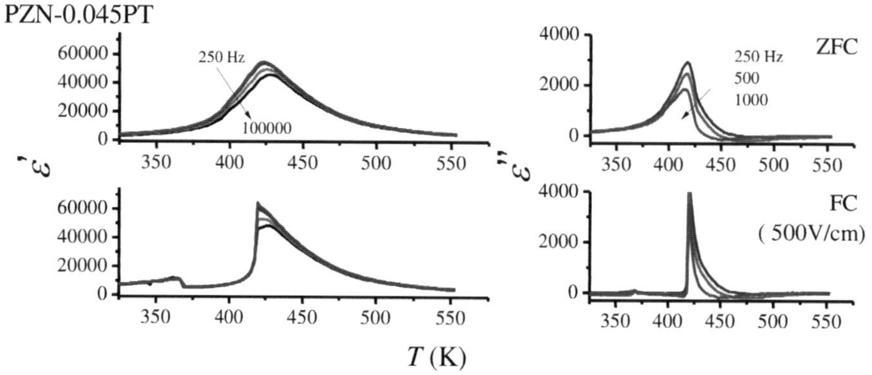

Figure 2. Temperature dependence of the real and imaginary parts of complex dielectric constant, $\varepsilon = \varepsilon' - j\varepsilon''$, obtained in a PZN-0.045PT single crystal on ZFC and FC under E= 500V/cm along [100]-direction.

electric field. On the other hand, marked changes are seen below the ferroelectric phase transition temperature. Even a small electric field of 100 kV/cm (~ $0.1E_{CEP}$) makes the diffuse phase transition sharp and two jumps and a plateau appear around 400 K in both ν_B and Γ. The plateau denotes the electric-field-induced tetragonal phase. With increasing the electric field, the tetragonal phase is stabilized as evidenced by the broadening plateaus. Despite the PZN–0.045PT approaches the critical end point by applying electric field, the amount of changes in ν_B and Γ at the phase transition does not enhanced, which must be attributed to the dynamics up to the GHz range.

3. 2 Dielectric properties

The dielectric properties probing lower frequency dynamics are shown in Figure 2. In common with the LA phonon, the tetragonal phase is induced by an electric field. The two peak temperatures in ε' under an electric field (360 K and 425 K) are in good agreement with two anomaly temperatures in the LA phonon, however, the peak temperature in ε' (421 K at 250 Hz) on zero-field cooling is different from the anomaly temperature in the LA phonon (403 K). It is because the complex elastic stiffness constant c_{11}^* is connected with ε' at 0 Hz when we assume that acoustic anomaly is a result of single relaxation process at the GHz range [11, 22].

As expected from the critical phenomena, the low frequency ε' is enhanced: a peak ε'at 250 Hz increases from 55000 to 65000, while ε'at 100 kHz increases from 45000 to 48000 by applying 500 V/cm (~ $0.5E_{CEP}$). The increase in ε'indicates that ε'at only low frequency is enhanced markedly at the critical end point. This idea is consistent with the LA phonon at the GHz range shown in Figure 1. Because the Brillouin scattering probes around the 42 GHz, little enhancement in $\varepsilon'$is detected through the strain-polarization coupling. In other words, the enhanced relaxation is located below the GHz range.

3. 3 Effect of electric field on phase transitions

Effect of an electric field on diffuse phase transitions in a relaxor ferroelectric PZN-0.045PT single crystal is studied through dielectric properties and a longitudinal acoustic (LA) phonon. Compared with the diffuse phase transition observed in zero-field cooling in PZN-0.045PT, it shows two sharp phase transitions under field cooling. According to our previous work on thermal hysteresis in PZN-0.07PT [12], the present data is analyzed.

From a mesoscopic point of view, the phase transition diffuseness is attributed to inhomogeneous structure from the static PNRs, nanodomains, and microdomains. When the relaxor ferroelectrics are cooled across T_C from the highest temperature, the size of the inhomogeneity just below T_C must be distributed, which smears the phase transition. When the relaxor ferroelectrics is cooled with an

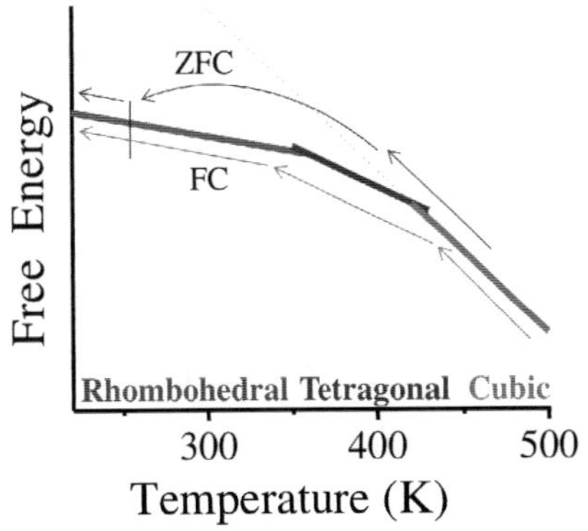

Figure 3. Schematic of free energies predicted for PZN–0.045PT around the phase transition temperatures to explain the observed difference between the data on ZFC and FC. The energies of the three phases are so close that supercooling can easily occur. On the other hand, the application of the electric field stabilizes each phase.

electric field, the static PNRs, nanodomains, and microdomains unite with others, and thus the ferroelectric domains grow in size and in polarization. Therefore, the complex inhomogeneous structure is eliminated, making the phase transitions sharp. The effect of inhomogeneity on a phase transition behavior is not so large in normal ferroelectrics. However, in the case of PZN–0.045PT, the cubic, tetragonal, and rhombohedral phases are in such competition around 400 K, in addition to the inhomogeneous structure, that the inhomogeneity can have a greater influence on the phases.

In general, more than one minimum in the free energy is thought to explain a first-order phase transition, but in the present case, at least three phases–cubic, tetragonal, and rhombohedral–should be taken into account as shown Figure 3. These phases adopt similar free energies around the phase transitions; as a result, nonequilibrium states can be easily achieved through supercooling, leading to the appearance of the drastic change by applying an electric field. When the sample is cooled without an electric field from the high temperature, the paraelectric cubic phase transforms to the supercooled ferroelectric rhombohedral state. On the field cooling, the electric field enhances the relaxation from the nonequilibrium polymorphic states to the equilibrium phase. Thus, each phase transition can be clearly distinguished.

4. Conclusion

The LA phonon and dielectric properties under an electric field in a PZN–0.045PT single crystal were measured to investigate the phase transition in relaxor ferroelectrics. The diffuse phase transition in PZN–0.045PT becomes sharp by applying the electric field during the phase transition, which is the result of reduced inhomogeneity in the complex relaxor system by an electric field. An external electric field grows the static PNRs and microdomains into larger ferroelectric domains.

By the complementary measurements of the LA phonon and dielectric properties, we discuss the phase transition dynamics in a broad frequency range up to the GHz range. The results indicate that enhanced dielectric property due to the applied electric field appears only in a low frequency region, which is necessary to be confirmed by measuring at various conditions in the future work.

Acknowledgments
This work was supported by JSPS KAKENHI Grant Number 30150981.

References

[1] Blinc R 2011 *Advanced Ferroelectricity* (New York: Oxford Univ Press)
[2] Bokov A A and Ye Z G 2006 *J. Mater. Sci.* **41** 31
[3] Dagotto E 2005 *New J. Phys.* **7** 67

[4] Tadaki T, Otsuka K and Shimizu K 1988 *Annu. Rev. Mater. Sci.* **18** 25
[5] Ko J-H, Kim D H and Kojima S 2008 *Phys. Rev. B* **77** 104110
[6] Ko J-H, Kim D H, Tsukada S, Kojima S, Bokov A A and Ye Z-G 2010 *Phys. Rev. B* **82** 104110
[7] Kojima S and Tsukada S 2010 *Ferroelectrics* **405** 32
[8] Tsukada S, Ike Y, Kano J, Sekiya T, Shimojo Y, Wang R P and Kojima S 2006 *Appl. Phys. Lett.* **89** 212903
[9] Tsukada S, Ike Y, Kano J, Sekiya T, Shimojo Y, Wang R and Kojima S 2008 *J. Phys. Soc. Jpn.* **77** 033707
[10] Tsukada S and Kojima S 2008 *Phys. Rev. B* **78** 144106
[11] Tsukada S, Hidaka Y, Kojima S, Bokov A A and Ye Z-G 2013 *Phys. Rev. B* **87** 014101
[12] Tsukada S, Kim T H and Kojima S 2013 *APL Materials* **1** 032114
[13] Kutnjak Z, Petzelt J and Blinc R 2006 *Nature* **441** 956
[14] Kutnjak Z, Blinc R and Ishibashi Y 2007 *Phys. Rev. B* **76** 104102
[15] Iwata M, Sakakibara K, Aoyagi R, Maeda M and Ishibashi Y 2009 *J. Ceram. Soc. Jpn.* **117** 954
[16] Iwata M, Sakakibara K, Aoyagi R, Maeda M and Ishibashi Y 2010 *Ferroelectrics* **405** 39
[17] Ohwada K, Hirota K, Rehrig P W, Fujii Y and Shirane G 2003 *Phys. Rev. B* **67** 094111
[18] Kim D H, Ko J H, Feng C D and Kojima S 2005 *Appl. Phys. Lett.* **87** 072908
[19] Akishige Y and Kamishina Y 1992 *J. Phys. Soc. Jpn.* **61** 4589
[20] Ko J H, Kim T H, Tsukada S, Kojima S and Koo T Y 2011 *Ferroelectrics* **420** 66
[21] Rehwald W 1973 *Adv. Phys.* **22** 721
[22] Tsukada S, Hiraki Y, Akishige Y and Kojima S 2009 *Phys. Rev. B* **80** 012102

IOP Conf. Series: Materials Science and Engineering **54** (2014) 012006 doi:10.1088/1757-899X/54/1/012006

Progress of ultrafast terahertz time-domain spectroscopy: Raman inactive soft mode in quantum paraelectric $SrTiO_3$

Tatsuya Mori*, Hikaru Igawa, and Seiji Kojima

Division of Materials Science, University of Tsukuba, Tennodai 1-1-1, Tsukuba, Ibaraki 305-8573, Japan

*E-mail: mori@ims.tsukuba.ac.jp

Abstract. We have measured the complex dielectric constants of quantum paraelectric $SrTiO_3$ (STO) bulk single crystals by using the ultrafast and broadband terahertz time-domain spectroscopy (THz-TDS) with the reflection configuration. The observed complex dielectric dispersion consists of the Raman inactive TO1 soft mode and is well reproduced by damped harmonic oscillator model where the resonant frequency and the damping constant are 2.75 THz and 0.63 THz, respectively, in good agreement with the previous reports. The ultrafast THz-TDS used in this study employs the technique of high-speed asynchronous optical sampling and Cherenkov type THz generator. The sampling time of one scan is about 10 milliseconds and available frequency of the THz source is from 0.5 to 7 THz. The reflection configuration combined with the ultrafast system enabled to detect the dispersion of a STO bulk sample which has high permittivity and opaque in the THz region.

1. Introduction

Terahertz time-domain spectroscopy (THz-TDS) which is one kind of far-infrared measurement technique is a powerful tool for THz material science [1-4]. THz-TDS records a time-domain waveform of the electric field of the THz electromagnetic wave. Since the measured THz waveforms have information of both amplitude and phase of the electric field, the complex optical constants can be determined directly without the Kramers-Kronig relation.

One of disadvantage of THz-TDS is a long measurement time due to the mechanical delay stage. The measurement time of the THz pulse shape by conventional THz-TDS is about several minutes and this long measurement time causes inaccuracy in experiment. But recently, ultrafast THz-TDS measurement technique has been established using high-speed asynchronous optical sampling (AOS) technique [5]. This technique uses two femtosecond lasers in substitution for mechanical delay stage, and the repetition rates of those lasers are slightly modulated. The AOS technique drastically decreases measurement time down to about 10 milliseconds for one shot of THz pulse. In addition to the ultra-fast scan of THz-TDS, we can use widely tunable monochromatic Cherenkov phase-matched THz wave generator, which was developed using nonlinear optic crystals of ferroelectric MgO doped $LiNbO_3$ [6].

In condensed matter physics, terahertz photon energy (1 THz = 4.14 meV) corresponds to minimal energy comparing to Fermi energy in electronic structure and corresponds to low-frequency optical phonon energy in phonon structure. Therefore, THz light is appropriate for the detection of such electronic and phonon structure [2, 3]. But THz light is opaque for the metallic and ferroelectric materials due to the high conductivity or high permittivity. So that THz-TDS with reflection configuration having large advantage for conventional FT-IR spectroscopy is strongly required for those materials.

$SrTiO_3$ (STO) is well known as a quantum paraelectric material, and has perovskite structure (space group: $Pm\bar{3}m$) at room temperature [4, 7-10]. STO has infrared active but Raman inactive TO1 soft phonon mode in around 3 THz. Due to this soft mode, STO is opaque and has high reflectivity in the THz region. Therefore, the soft mode was studied by Raman scattering by the application of electric or stress fields to suppress the centrosymmetry. While, the study of soft mode without any external fields is very important, because structural phase transitions are very sensitive to external fields.

In this study, we measured TO1 soft mode of a STO bulk single crystal using the ultrafast and broadband THz-TDS with reflection configuration, in addition to the transmission measurement below 2 THz. The combination of two configuration measurements enables the accurate determination of the frequency dependent complex dielectric constant directly without any assumptions.

2. Experimental

The (100) plate of STO single crystal from Furuuchi Chemical, whose size is $10 \times 10 \times 0.5$ mm^3 with two optically polished surfaces, was used in this study. The reflectivity spectra were measured by ultrafast THz-TDS (TAS7500SU, Advantest Co.) using a Cherenkov type THz generator and the high-speed AOS technique at room temperature. The reflection configuration is near normal incident and incident angle is about 12°. As a reference, a mirror polished aluminium plate was used. As shown in figure 1, the reflectivity reference THz spectrum has a frequency range of from 0.5 to 7 THz.

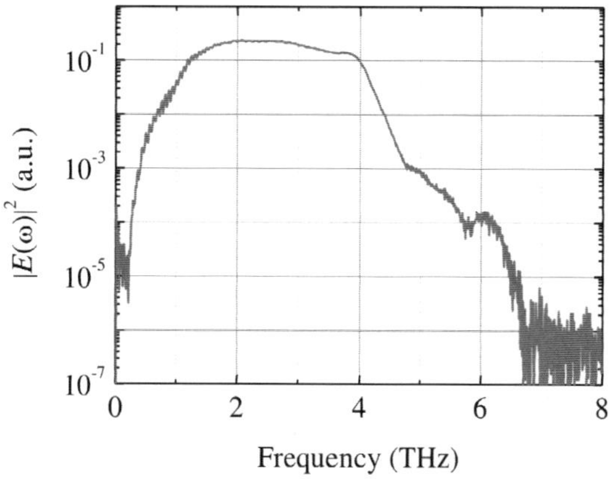

Figure 1. Power spectrum of the broad band THz pulse reflected from an aluminum reference mirror. An available frequency range is from about 0.5 to 7 THz. For convenience, 1 THz = 33.3 cm^{-1} = 4.14 meV = 48 K.

3. Results and Discussion

Figures 2(a) and (b) show the measured reflectivity and phase shift of a STO single crystal at room temperature. The reflectivity increases toward higher frequency and shows strong increase especially around 2.5 THz. Those reflectivity behaviors are consistent with the reflectivity spectrum of damped harmonic oscillator (DHO) and the observed mode is assigned to TO1 soft mode.

The measured phase shift, which is shown at figure 2(b) by a red solid line, increases toward higher frequency corresponding to reflectivity behavior. In the reflectivity measurement, the reference and

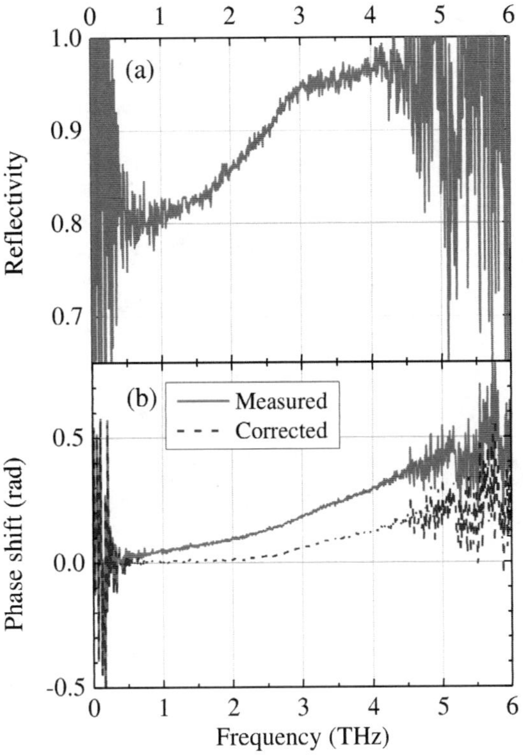

Figure 2. (a) Reflectivity and (b) phase shift of STO. Solid lines (red) and a dashed line (blue) are raw and corrected data, respectively.

the sample have position discrepancy caused by technical inaccuracy of the sample setting on the sample stage. The discrepancy causes incorrect determination of the complex optical constants. To eliminate the position discrepancy between an aluminium reference and the STO sample, we correct the measured phase shift by the following procedure. First, we measured the STO sample by transmission configuration and extracted the complex dielectric constants. The results of the transmission measurements are shown in figures 3(a) and (b) by open squares. Since the results of transmission THz-TDS measurements for STO bulk sample have high accuracy as shown in figure 3, we fitted the results of the reflectivity measurement by using the transmission data. The measured waveform of the sample is shifted with 6.7 femtoseconds toward the reference peak direction and finally the corrected phase shift is obtained by Fourier transform of the shifted time-domain waveform.

Using the reflectivity and the corrected phase shift shown in figure 2(b) by a blue dashed line, we extracted the complex refractive index by the following equation,

$$\frac{E_{sam}(\omega)}{E_{ref}(\omega)} = \frac{1-n}{1+n}, \tag{1}$$

where $n = n' + i \cdot n''$ is the complex refractive index. $E_{sam}(\omega)$ and $E_{ref}(\omega)$ are the complex amplitude spectra of the THz pulse reflected from the sample and the aluminium reference mirror, respectively. The extracted complex refractive index is converted to the complex dielectric constant by the following relation,

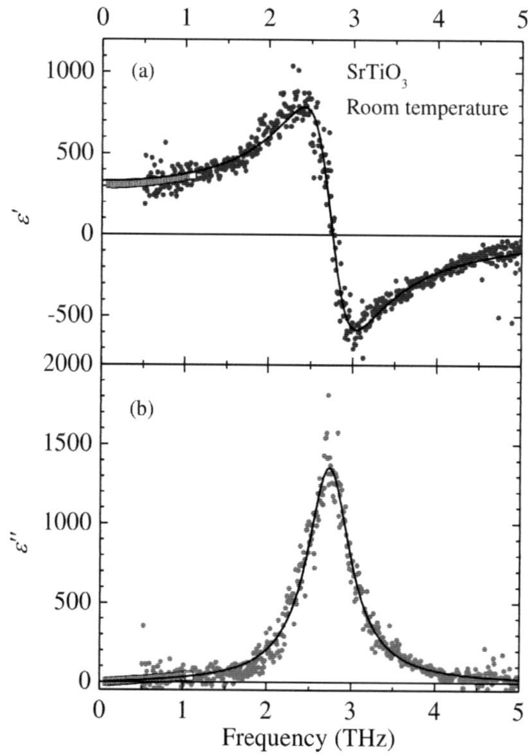

Figure 3. (a) Real and (b) imaginary parts of complex dielectric constant of STO single crystals at room temperature. Filled circles are the results of the reflectivity measurements. Open squares are extracted from the transmission measurements. Solid lines are the fitting results of a single DHO model.

Table 1. Fitting parameters for TO1 soft mode of STO at room temperature.

ε_{dc}	ω_1 (THz)	S_1	Γ_1 (THz)
331	2.75	308	0.63

$$\varepsilon = n^2, \quad (2)$$

where $\varepsilon = \varepsilon' + i \cdot \varepsilon''$ is the complex dielectric constant.

Figures 3(a) and (b) show the real and imaginary parts of complex dielectric constant, respectively. The entire structure of TO1 soft mode is clearly observed and the behaviour is consistent with frequency dependence of DHO model. The ε' crosses zero and the ε'' peaks at 2.7 THz indicating the resonant frequency of the soft mode. The experimental results are fitted by a single DHO model as follows,

$$\varepsilon(\omega) = \varepsilon_\infty + \frac{S_1\omega_1^2}{\omega_1^2 - \omega^2 - i\omega\Gamma_1}, \quad (3)$$

IOP Conf. Series: Materials Science and Engineering **54** (2014) 012006 doi:10.1088/1757-899X/54/1/012006

with $\varepsilon_\infty = \varepsilon_{dc} - S_1$, where ε_{dc} is static dielectric constant, ω_1, S_1 and Γ_1 are resonant frequency, oscillator strength and damping constant of TO1 soft phonon mode, respectively. The parameters obtained by the fitting are listed in Table I. The resonant frequency and the damping constant are 2.75 THz and 0.63 THz, respectively and those results are consistent with previous studies [4, 7-10].

4. Conclusion

We successfully obtained the complex dielectric constant of a quantum paraelectric STO single crystal using ultrafast and broadband THz-TDS with reflection configuration. The Raman inactive TO1 soft mode behavior is well reproduced by DHO model and the obtained fitting parameters are consistent with previous studies [4, 7-10]. The measurement of temperature dependence is in progress to clarify the soft mode behavior of quantum paraelectric STO, which can be related to the Barrett's relation of dielectric permittivity [11]. This ultrafast and reflectivity measurement technique will be applicable to various bulk materials. The combination of reflectivity and transmission configuration measurements enables us to determine directly the reliable complex optical constants of opaque materials in the THz region without any assumptions.

Acknowledgment

Authors are very thankful for the technical support on ultrafast THz-TDS measurements to Sendai Office, Advantest Corporation. This work was partially supported by Grants-in-Aid for Young Scientists (B)(24740194) and the Murata Science Foundation.

References

[1] Kojima S, Tsumura N, Wada Takeda M and Nishizawa S 2003 *Phys. Rev. B* **67** 035102
[2] Mori T, Nicol E J, Shiizuka S, Kuniyasu K, Nojima T and Carbotte J P 2008 *Phys. Rev. B* **77** 174515
[3] Mori T, Iwamoto K, Kushibiki S, Honda H, Matsumoto H, Toyota N, Avila M A, Suekuni K and Takabatake T 2011 *Phys. Rev. Lett.* **106** 015501
[4] Matsumoto N, Fujii T, Kageyama K, Takagi H, Nagashima T and Hangyo M 2009 *Jpn. J. Appl. Phys.* **48** 09KC11
[5] Bartels A, Hudert F, Janke C, Dekorsy T and Köhler K 2006 *Appl. Phys. Lett.* **88** 041117
[6] Shibuya T, Suizu K and Kawase K 2010 *Appl. Phys. Express* **3** 082201
[7] Spitzer W G, Miller R C, Kleinman D A and Howarth L. E. 1962 *Phys. Rev.* **126** 1710
[8] Barker A S and Tinkham M 1962 *Phys. Rev.* **125** 1527
[9] Misra M, Kotani K, Kawayama I, Murakami H and Tonouchi M 2005 *Appl. Phys. Lett.* **87** 182909
[10] Petzelt J, Ostapchunk T, Gregora I, Rychetský I, Hoffmann-Eifert S, Pronin A V, Yuzyuk Y, Gorshunov B P, Kamba S, Bovtun V, Pokorný J, Savinov M, Porokhonskyy V, Rafaja D, Vaněk P, Almeida A, Chaves M R, Volkov A A, Dressel M and Waser R 2001 *Phys. Rev. B* **64** 184111
[11] Barrett J H 1952 *Phys. Rev.* **86** 118

Structural and optical properties of CdSe nanostructures (nanoparticles, nanoparticle- and nanosheet-superlattices) fabricated using organic molecules as a template

Kiyoto Matsuishi[1], Akimasa Yuasa[1], Go Arai[1] and Tatsuo Mori[2]

[1] Division of Materials Science, Faculty of Pure and Applied Sciences, University of Tsukuba, Tennoudai 1-1-1, Tsukuba, Ibaraki 305-8573, Japan
[2] Ibaraki National College of Technology, Nakane 866, Hitatinaka, Ibaraki 312-8508, Japan

E-mail: kiyoto@bk.tsukuba.ac.jp

Abstract. Structural and optical properties of trioctylphosphine-oxide-capped CdSe nanoparticles in an wurtzite phase were investigated to obtain the size dependences of bulk modulus and deformation potential by X-ray diffraction and photoluminescence measurements under high pressure. The bulk modulus B [GPa] and the deformation potential a [eV] were described as a function of the diameter D [nm] of CdSe nanoparticles by $B = -110\times(1/D) + 53$ and $a = 5.13\times(1/D) - 2.30$ in the pressure range to 3 GPa. Structural and optical properties of three-dimensional CdSe nanoparticle-superlattice and two-dimensional CdSe nanosheet-superlattice were also studied at ambient pressure and under high pressure. Enhanced interaction between nanoparticles in CdSe nanoparticle-superlattices, compared with CdSe nanoparticles dispersed in solution, was observed in their optical properties. For two-dimensional CdSe nanosheet-superlattices, two excitonic bands whose energy difference was much larger than that of bulk CdSe crystals were observed in optical absorption spectra. The energy of splitting between the two excitonic bands increased linearly with pressure. The observations indicate that the crystal field effect on the band mixing is enhanced significantly in the nanosheet-superlattice structure.

1. Introduction

Semiconductor nanostructures (nanoparticles (NPs), nanorods (NRs) and nanosheets (NSs)) exhibit unique size-dependent optical properties that are significantly distinguished from its bulk state. Especially, CdSe-related nanostructures have drawn much interest in viewpoints of both physics and optical applications due to their easily-tunable high efficient luminescence in the visible range by changing the nanoscale dimension [1, 2]. However, to our best knowledge, there have been few experimental works on the size dependences of bulk modulus and deformation potential of CdSe nanostructures.

A self-organized method using organic molecules as a template to form organic-inorganic hybrid superlattice (SL) has been utilized to fabricate CdSe nanostructures with controlled dimensionality [3, 4]. However, optical and electronic properties of organic-CdSe hybrid SL have not been studied sufficiently and therefore have not been fully understood.

In this study, we have studied the size dependences of high pressure properties, especially the bulk modulus and the deformation potential of TOPO (trioctylphosphine oxide)-capped CdSe NPs ranging from 2.8 to 5.4 nm in diameter. In addition, we have investigated structural, optical and electronic properties of three-dimensional CdSe nanoparticle-superlattice (CdSe NP-SL) and two-dimensional CdSe nanosheet-superlattice (CdSe NS-SL) at ambient pressure and also under pressure to discuss effects of interaction between NPs and NSs on excitonic states by changing the distance between adjacent NPs and NSs, respectively, in the SL structures.

2. Experimental

TOPO-capped CdSe NPs with diameters *D* ranging from 2.8 to 5.4 nm were prepared by a conventional hot-injection method [5]. The sizes of the prepared CdSe NPs were determined by using the relationship between the NP diameter and the lowest excitation energy obtained from optical absorption spectra in an UV-visible region [6]. The surface modification on CdSe NPs was also made by altering TOPO by HDA (hexadecylamine) to explore the effects of the surface states on the bulk modulus and deformation potential. After carefully purifying TOPO-capped CdSe NPs, X-ray diffraction (XRD) and photoluminescence (PL) were measured under high pressure to 9 GPa using a diamond anvil cell (DAC) with a 4:1 methanol-ethanol mixture as a transmitting medium. High-pressure XRD measurements were carried out at the BL-18C beamline (wavelength: 0.6198A) at KEK-PF. PL spectra were measured under high pressure using the 442 or 325 nm line of a He-Cd laser as an excitation light source.

TOPO-capped CdSe NPs were arrayed three-dimensionally to form CdSe NP-SLs by a pressure-controlled solution-evaporation method as follows. TOPO-capped CdSe NPs were dispersed in the mixture of n-octane and 1-octanol (the ratio was 9:1). The dispersed solution was heated up to 80 °C, and then evacuated to 1 torr at the rate of 3 torr/min, and kept at 80 °C at 1 torr for two days to grow crystals of CdSe NP-SLs. SL structures were confirmed and their lattice parameters were determined by small angle X-ray scattering (SAXS; Rigaku, NANO-Viewer) for CdSe NP-SLs.

CdSe NS-SLs with alkylamine molecules acting as a barrier layer as well as a template were fabricated by a soft-colloidal template method [4]. Two-dimensional SL structures and their lattice parameters were examined by conversional XRD measurements.

Optical absorption and PL spectra of CdSe NP-SLs and CdSe NS-SLs were measured under hydrostatic pressure to 6 GPa using a DAC with either cyclohexene or a 4:1 methanol-ethanol mixture as a transmitting medium. Time-resolved PL measurements to obtain fluorescence life times were performed at ambient pressure using a streakscope with a spectrograph and a N_2 pulse laser (337 nm) for CdSe NP-SLs and a pico-seconds laser diode (375 nm) for CdSe NS-SLs as an excitation light source.

3. Results and discussion

3.1. Bulk modulus and deformation potential of CdSe NPs

High-pressure XRD patterns of TOPO-capped CdSe NPs exhibited an wurtzite structural phase below 3 GPa and recovered to the original ones after releasing to atmospheric

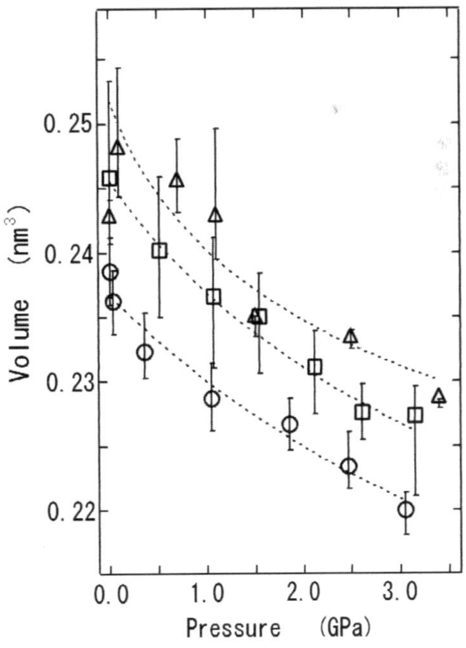

Figure 1. Calculated unit cell volumes of CdSe NPs with *D* = 2.8, 3.1 and 5.4 nm represented by Δ, □, ○, respectively, as a function of pressure. The dashed lines are the curves fitted to the data using equation (1).

Table 1. Experimentally-obtained parameters of CdSe NPs.

Diameter D [nm]	Bulk Modulus B [GPa]	Pressure Coefficient [meV/GPa]	Deformation Potential a [eV]
2.8	11.9	14.4	− 0.38
3.1	21.6	16	− 0.74
4.1		21.3	
5.4	29.4	23.5	−1.34

pressure. We obtained the unit cell volume V as a function of pressure P, as shown in figure 1, from high-pressure XRD data below 3 GPa, and then calculated the bulk modulus B for CdSe NPs with D = 2.8, 3.1 and 5.4 nm, as tabulated in table 1, by fitting V using the Murnaghan's equation of state;

$$P(V) = \frac{B}{B'}\left[\left(\frac{V_0}{V}\right)^{B'} - 1\right], \qquad (1)$$

where B' is the pressure derivative of B, dB/dP, and V_0 is the unit cell volume at atmospheric pressure. We found that the bulk modulus for CdSe NPs is much smaller than that of bulk CdSe crystals (B = 53.0 GPa [7]) and decreases as the size of NP decreases. We plotted B as a function of $1/D$ in figure 2 and found that B [GPa] decreases linearly with the inverse of D [nm], as expressed well by $B = -110\times(1/D) + 53$.

We observed a PL band originating from the excitonic (or HOMO-LUMO gap) emission which shifts linearly to higher energies with pressure in the low pressure region to 3 GPa. The pressure coefficient of the energy of the PL band is tabulated in table 1 and plotted as a function of $1/D$ in

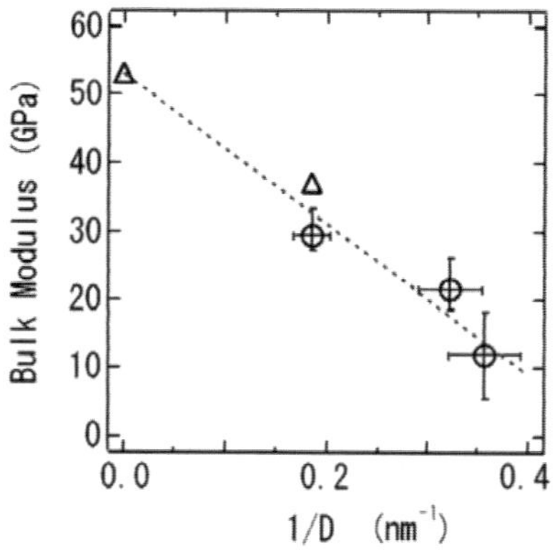

Figure 2. Bulk modulus as a function of $1/D$ for CdSe NPs. The open circles are from the present study, and the open triangles from Refs. [7, 8].

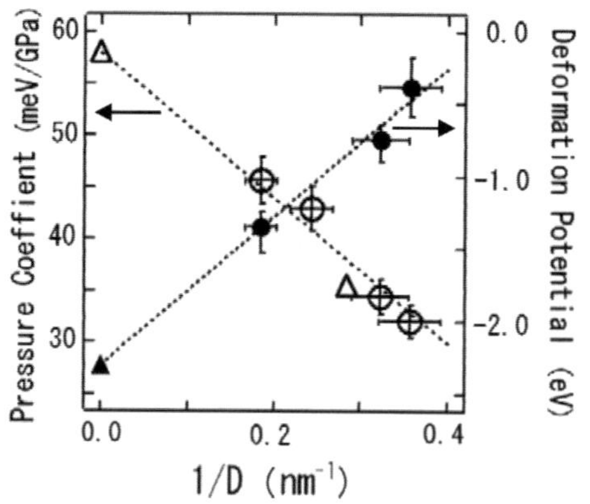

Figure 3. Pressure coefficient of PL peak energy and deformation potential as a function of $1/D$ for CdSe NPs. The open circles and triangles represent the pressure coefficients from the present study and from Refs. [7, 9], respectively. The solid circles and triangles represent the deformation potentials from the present study and from Ref. [10], respectively.

figure 3. The size dependence of the pressure coefficient of the HOMO-LUMO gap for CdSe NPs can be understood in the light of the enhancement of mixing of electronic states at off-Γ points to that at Γ point in the Brillouin zone of bulk CdSe crystals by decreasing the size of NP. We found also that the Stokes shift increases with pressure, e.g. by 6 meV/GPa for the sample of D = 2.8 nm.

We estimated the deformation potential a using the bulk modulus and the pressure coefficient of the energy of the PL band, as tabulated in table 1. We found that the absolute value of deformation potential for CdSe NPs is much smaller than that of bulk CdSe crystals (a = – 2.3 eV [10]) and decreases as the size of NP decreases. We plot a as a function of $1/D$ in figure 3 and find a linear relationship between a [eV] and $1/D$ [nm^{-1}], as expressed by $a = 5.13\times(1/D) - 2.30$. It should be pointed out that the linear dependence of deformation potential on $1/D$ did not change by the surface modification by HDA, even though the pressure coefficient of the energy of the PL band depended on the surface ligands.

3.2. Structural and optical properties of CdSe NP-SLs

TOPO-capped CdSe NPs with D = 2.3, 3.2, 3.7 and 4.2 nm were arrayed three-dimensionally by a pressure-controlled solution-evaporation method. We obtained small platelets of CdSe NP-SLs. The SAXS profiles shown in figure 4 exhibit broad, but appreciable, peaks originating from (1 1 1) and (2 2 2) diffraction peaks of simple cubic (SC) structures, providing confirmation of three-dimensional SL structures. We found that the lattice parameter of SC structure increased from 3.7 to 5.0 nm as D increased from 2.3 to 4.2 nm. The result implies that the distance between the nearest surfaces of two adjacent NPs, d_{NP}, was decreased from 1.4 to 0.8 nm as D increased from 2.3 to 4.2 nm.

After dispersed-CdSe NPs were arrayed three-dimensionally for CdSe NP-SL, the excitonic emission band in PL spectra was red-shifted and became asymmetric in lineshape, as shown in figure 5 for D = 3.2, 3.7 and 4.2 nm. Here, the change in the energy of the PL band with D for dispersed-CdSe NPs can be well explained by a quantum confinement model [11]. It should be noted that the red-shift of the PL band after arraying to three-dimensional SL structure depends on d_{NP}. Roughly speaking, the red-shift becomes larger as d_{NP} decreases. The observation suggests that the energy transfer between NPs occurs in the SL structures and is promoted as d_{NP} decreases.

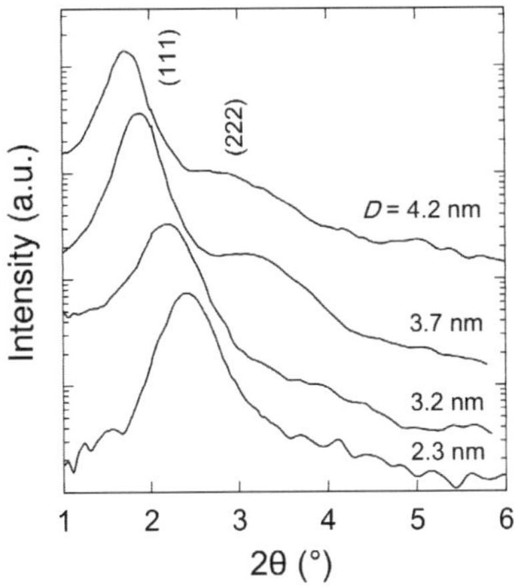

Figure 4. Small angle X-ray scattering profiles of CdSe NP-SLs with D = 2.3, 3.2, 3.7 and 4.2 nm. The broad, but appreciable, peaks can be indexes by a simple cubic SL structure.

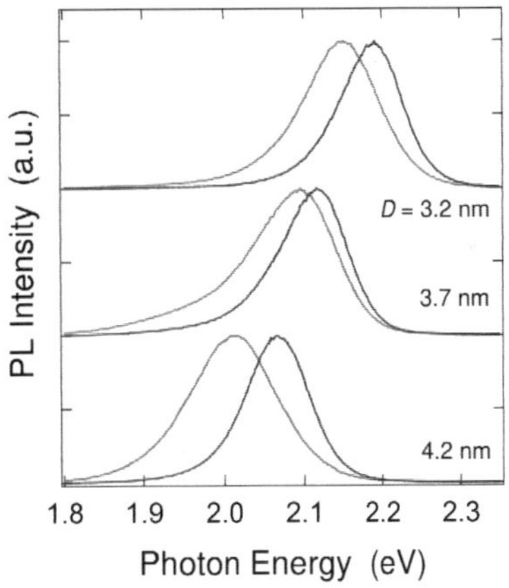

Figure 5. PL spectra of CdSe NP-SLs with D = 3.2, 3.7 and 4.2 nm (red color) in comparison of those of dispersed-CdSe NPs before arraying (blue color). The 442 nm line (2.81 eV) of He-Cd laser was used as an excitation light source.

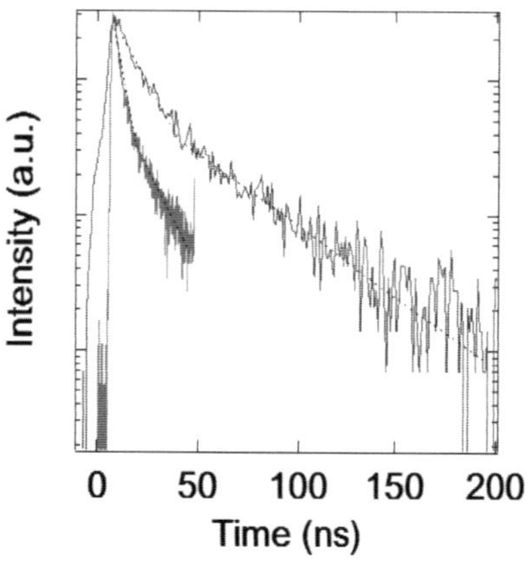

Figure 6. Time-resolved PL intensity as a function of decay time for dispersed-CdSe NP (blue color) and CdSe NP-SL (red color) with $D = 3.5$ nm.

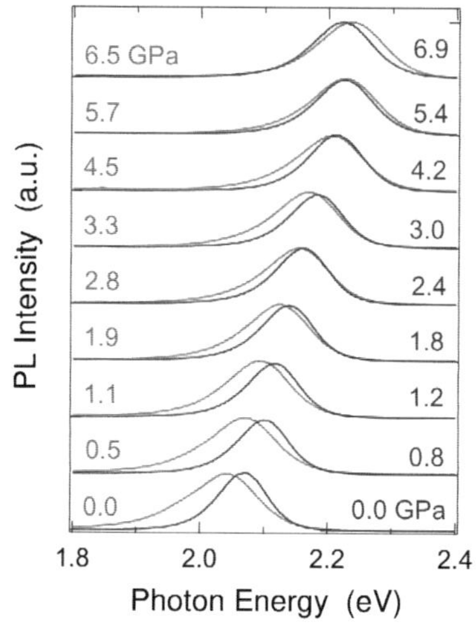

Figure 7. Change in PL spectra with pressure for dispersed-CdSe NP (blue color) and CdSe NP-SL (red color) with $D = 4.2$ nm. The 442 nm line (2.81 eV) of He-Cd laser was used as an excitation light source.

Figure 6 shows the time-resolved PL intensity as a function of decay time for the excitonic emission band of dispersed-CdSe NP and CdSe NP-SL with $D = 3.5$ nm. We analyzed the data by curve-fitting to a first-order kinetics equation with two lifetimes (i.e. two exponential decay components) to evaluate an effect of interaction between NPs on the excitonic state. From the curve-fitting, the lifetimes were estimated to be 10.2 and 41.3 ns for dispersed-CdSe NP and 2.4 and 15.2 ns for CdSe NP-SL. The result implies that the exciton life times were shorten by one-fourth for the fast component and by one-third for the slow component after dispersed-CdSe NPs were arrayed to the SL structure. This is another evidence of enhanced interaction between NPs in CdSe NP-SLs as a result of the formation of SL structure.

Figure 7 shows the pressure dependence of the excitonic emission band in PL spectra for dispersed-CdSe NP and CdSe NP-SL with $D = 4.2$ nm. With applying hydrostatic pressure, the PL peak of CdSe NP-SL increases in energy and becomes symmetric in lineshape. It is noteworthy that it becomes eventually almost identical in appearance with that of dispersed-NP around 6 GPa. We found that the peak energy of the excitonic emission band shifts linearly to higher energies with pressure to 3 GPa for both dispersed-CdSe NP and CdSe NP-SL. However, the pressure coefficient for CdSe NP-SL is larger by 10 % than that for dispersed-CdSe NP.

3.3. Structural and optical properties of CdSe NS-SLs

Figure 8 shows the XRD patterns in the low angle region ($2\theta = 3 - 15°$) for CdSe NS-SLs with hexylamine, octylamine, dodecylamine and oleylamine. The (0 0 l) diffraction peaks ($l = 2, 3$) were clearly observed, indicating a two-dimensional SL structure with the c-axis (the stacking direction of CdSe nanosheets and alkylamine layers) normal to the plane of flakes of sample. The observed diffraction peaks shift to higher angles with longer alkylamines used for the barrier layer as well as for the template. The lattice constant of the two-dimensional SL structure, i.e. the distance between adjacent CdSe nanosheets, was estimated and plotted in figure 9 as a function of the number of carbon atoms of the alkylamine used for CdSe NS-SLs.

IOP Conf. Series: Materials Science and Engineering **54** (2014) 012007 doi:10.1088/1757-899X/54/1/012007

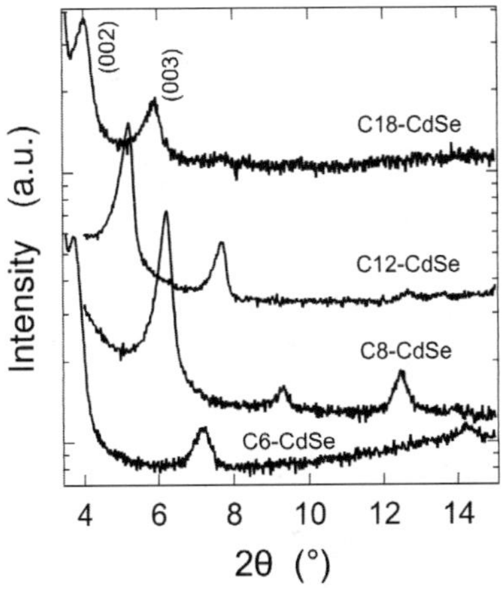

Figure 8. X-ray diffraction patterns of CdSe NS-SLs with hexylamine (C6-CdSe), octylamine (C8-CdSe), dodecylamine (C12-CdSe), oleylamine (C18-CdSe) in the low angle region. The Cu-Kα line was used as an X-ray source.

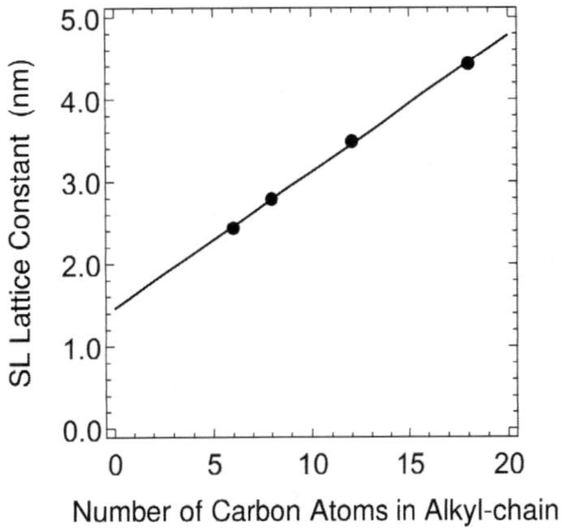

Figure 9. Lattice constant of two-dimensional SL structure (i.e. distance between adjacent CdSe nanosheets) as a function of the number of carbon atoms of the alkylamine acting as the barrier layer as well as the template for CdSe NS-SLs.

We found that the distance between adjacent CdSe nanosheets in CdSe NS-SL varied linearly with the number of carbon atoms in the alkylamine. By interpolating the linear relationship to zero in the number of carbon atoms in figure 9, we estimated the thickness of a CdSe nanosheet to be 1.4 nm (corresponding to about 7 atomic layers) which is consistent with the result of high-resolution transmission electron microscopy observation of CdSe NS-SL with octylamine by Son and his coworkers [4]. Consequently, the thickness of an alkylamine layer (i.e. barrier layer) was found to vary from 1.0 to 3.1 nm by using different alkylamines in order of hexylamine, octylamine, dodecylamine, oleylamine. Our XRD data in the high angle region ($2\theta = 20 - 50°$) revealed that the CdSe nanosheets have a slightly-distorted wurtzite structure. A previous work on selected-area electron diffraction measurements reported that the *c* axis of the wurtzite structure is parallel and the *a* axis perpendicular to the nanosheet [4]. Schematic structural image is sketched in figure 10 in aid of the findings by calculation using density fluctuation theory [4].

Optical absorption spectra of CdSe NS-SL with octylamine at ambient pressure and under high pressure are shown in figure 11. They exhibit two distinct sharp peaks originating from two excitonic states, 1_{hh}-1_e and 1_{lh}-1_e which are associated with heavy and light holes, respectively. The energy of splitting between the two peaks corresponds to the difference in energy between the heavy hole and the light hole in the valence band. We found that both 1_{hh}-1_e and 1_{lh}-1_e shift slightly to a higher energy with an increase in the thickness of barrier layer (i.e. the thickness of alkylamine layer) due to dielectric confinement of excitons. The energy of splitting between 1_{hh}-1_e and 1_{lh}-1_e at ambient pressure for CdSe NS-SL with octylamine is 186 meV at 5 K which is about 8 times larger than the energy difference between heavy and light holes at the Γ point in the valence band of bulk CdSe crystals (25 meV at 4.2 K). The significant increase in the energy of splitting between heavy and light holes in CdSe NS-SL compared with that of bulk CdSe crystals can be explained with an increase in the crystal field effect by the quasi-cubic model of band mixing for wurtzite crystals [12] in addition to an exciton confinement effect in CdSe nanosheets.

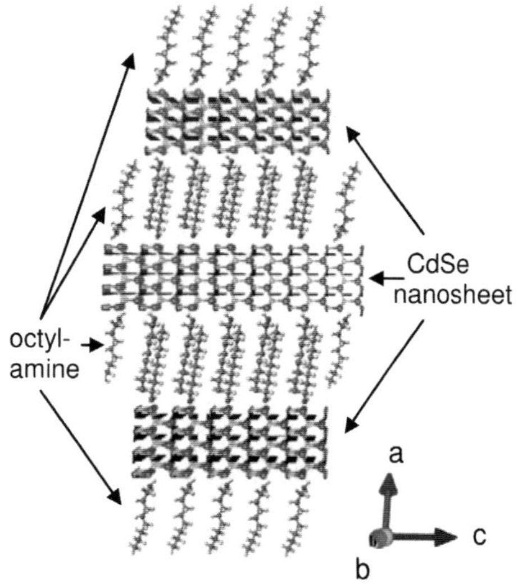

Figure 10. Schematic structural image of CdSs NS-SL with octylamine. Pink and green solid circles represent Cd and Se atoms, respectively, in CdSe nanosheets.

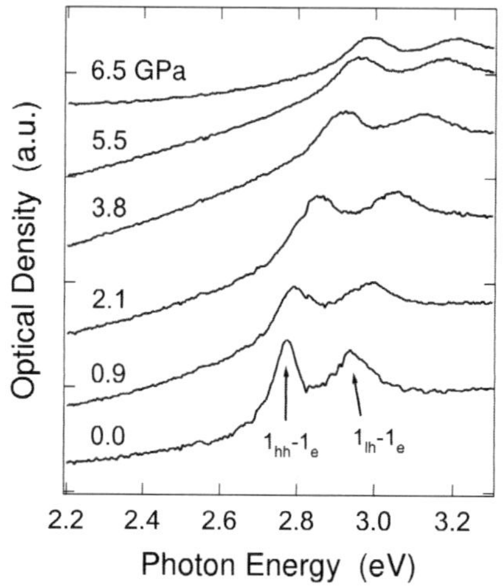

Figure 11. Change in optical absorption spectra with pressure for CdSe NS-SL with octylamine.

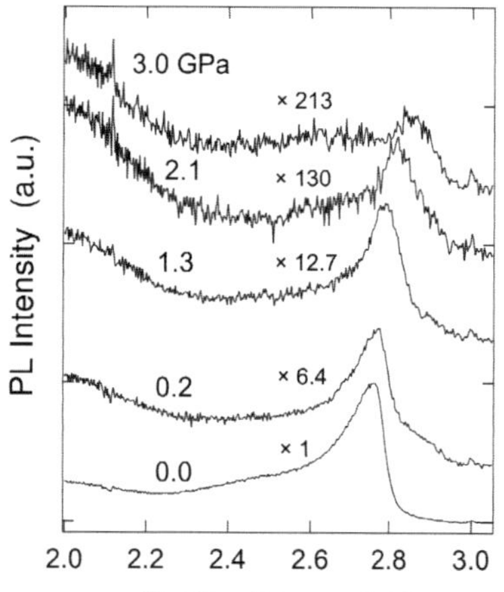

Figure 12. Change in PL spectra with pressure for CdSe NS-SL with octylamine. The 325 nm line (3.81 eV) of He-Cd laser was used as an excitation light source.

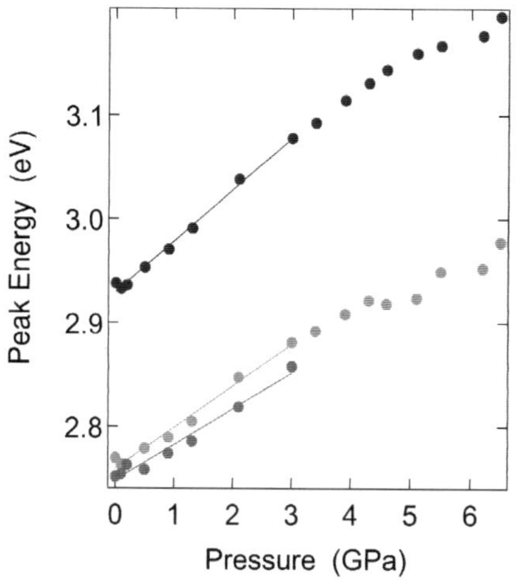

Figure 13. Pressure dependences of energies of two optical absorption peaks, 1_{hh}-1_e (green solid circles) and 1_{lh}-1_e (blue solid circles), and PL peak (red solid circles) for CdSe NS-SL with octylamine.

As shown in figure 11, both the absorption peaks of 1_{hh}-1_e and 1_{lh}-1_e shift to higher energies with applying pressure to 6.5 GPa. It should be noted that any indications of a phase transition can be realized in the pressure dependence of optical absorption spectra, while bulk CdSe crystals exhibit a structural phase transition from an wruzite to a rock salt structure at 2.7 GPa, which is accompanied by a significant change in an optical absorption spectra due to the direct-indirect band structural change. The pressure dependence of PL spectra of CdSe NS-SL with octylamine is shown in figure 12. The PL peak originating from the 1_{hh}-1_e exciton emission appears at 2.75 eV at ambient pressure and shifts to higher energies with pressure. We plot the energies of the two optical absorption peaks and the PL peak as a function of pressure in figure 13. All of them shift linearly with pressure to 3GPa. The pressure coefficients are 40 and 50 meV/GPa for the 1_{hh}-1_e and 1_{lh}-1_e absorption peaks, respectively, and 34 meV/GPa for the 1_{hh}-1_e exciton emission peak. The energy of splitting between 1_{hh}-1_e and 1_{lh}-1_e absorption peaks increases linearly with pressure and reaches 225 meV at 6 GPa. The observation indicates that the crystal field effect is enhanced significantly by applying pressure in CdSe nanosheets. Another interesting finding is that the Stokes shift, i.e. the difference in energy between the absorption and emission peaks of 1_{hh}-1_e exciton, increases linearly with pressure by 6 meV/GPa, suggesting that electron-lattice (or exciton-lattice) interaction is also enhanced with applying pressure in CdSe nanosheets. We have found that the fluorescence life time of 1_{hh}-1_e exciton emission band for CdSe NS-SLs decreases with increasing the thickness of alkylamine layers (i.e. barrier layer) by using different alkylamines, providing information about interaction of excitons between nanosheets. The details will be reported elsewhere.

4. Conclusion

We have studied the structural and electronic states of TOPO-capped CdSe NPs ranging from 2.8 to 5.4 nm in diameter under high pressure to explore the size dependences of bulk modulus and deformation potential. We found that a linear relationship between the bulk modulus and the size of NP as well as between the deformation potential and the size of NP below 3 GPa. We have also investigated the structural and optical properties of three-dimensional CdSe NP-SLs at ambient pressure and under high pressure to discuss effects of interaction between NPs on the excitonic states by changing the distance between adjacent NPs. Evidence of enhanced interaction between NPs in CdSe NP-SLs was obtained as a result of the formation of SL structure. For two-dimensional CdSe NS-SLs, two excitonic states, 1_{hh}-1_e and 1_{lh}-1_e were observed. The energy of splitting between heavy and light holes was much larger than that of bulk CdSe crystals, and increased linearly with pressure, indicating that the crystal field effect on the band mixing of wruzite structure is significantly enhanced in the NS-SL structure.

Acknowledgments
This work was supported in part by the Grant-in-Aid for Scientific Research from the Ministry of Education, Culture, Sports, Science and Technology in Japan. The high-pressure Synchrotron X-ray diffraction measurements were performed under proposal No. 2006G271 of the Photon Factory, KEK.

References

[1] Brus L 1986 *J. Phys. Chem.* **90** 2555
[2] Hoheisel W, Colvin V L, Johnson C S and Alivisatos A P 1994 *J. Chem. Phys.* **101** 8455
[3] Kagan C R, Murray C B and Bawendi M G 1996 *Phys. Rev.* B **54** 8633
[4] Son J S, *et al* 2009 *Angew. Chem. Int. Ed.* **48** 6861
[5] Murray C B, Norris D J and Bawendi M G 1993 *J. Am. Chem. Soc.* **115** 8706
[6] Yu W W, Qu L, Guo W and Peng X 2003 *Chem. Mater.* **15** 2854
[7] Wei S H and Zunger A 1999 *Phys. Rev.* B **60** 5404
[8] Tolbert S H and Alivisatos A P 1995 *J. Chem. Phys.* **102** 4642
[9] Zhao Z, Zeng J, Ding Z, Wang X and Hou J 2007 *J. Appl. Phys.* **102** 053509
[10] Shan W, Walukiewicz W, Ager III J W, Yu K M and Wu J 2004 *App. Phys. Lett.* **84** 67
[11] Kayanuma Y 1988 *Phys. Rev.* B **38** 9797
[12] Lew Yan Voon L C, Willatzen M, Cardona M and Christensen N E 1996 *Phys. Rev.* B **53** 10703

Sol–gel synthesis of nanosized λ-Ti_3O_5 crystals

T Nasu[1], H Tokoro,[1,2,3,*] K Tanaka[1], F Hakoe[1], A Namai[1] and S Ohkoshi[1,4,*]

[1] Department of Chemistry, School of Science, The University of Tokyo, 7-3-1 Hongo, Bunkyo-ku, Tokyo 113-0033, Japan
[2] Division of Materials Science, Faculty of Pure and Applied Sciences, University of Tsukuba, 1-1-1 Tennodai, Tsukuba, Ibaraki 305-8577, Japan
[3] NEXT, JSPS, 8 Ichibancho, Chiyoda-ku, Tokyo 102-8472, Japan
[4] CREST, JST, K's Gobancho, 7 Gobancho, Chiyoda-ku, Tokyo 102-0076, Japan

tokoro@ims.tsukuba.ac.jp, ohkoshi@chem.s.u-tokyo.ac.jp

Abstract. In this study, we show a synthesis of λ-Ti_3O_5 nanocrystals dispersed in silica by sol–gel method. The X-ray diffraction measurements, Rietveld analyses, and transmission electron microscope images of the obtained samples showed that tuning the sintering temperature in the synthesis process can control the size of the λ-Ti_3O_5 nanocrystals, i.e., 8±2 nm (1123°C; sample **1**), 9±3 nm (1133°C; **2**), 9±2 nm (1143°C; **3**), 10±3 nm (1153°C; **4**), 11±4 nm (1163°C; **5**), 13±4 nm (1173°C; **6**), 25±12 nm (1200°C; **7**), and 36±15 nm (1250°C; **8**), whereas adjusting the hydrogen flow rate can tune the oxidation-reduction state of the sample without apparent change in the crystal size. At the lowest sintering temperature of 1123°C, the smallest λ-Ti_3O_5 nanocrystals of 8 nm in size were produced.

1. Introduction

Photo-induced phase transition materials such as chalcogenides are attractive for both fundamental and applied research [1–4]. Chalcogenides are used as the recording phase-change material in storage media, e.g., Digital versatile discs (DVDs) and Blu-ray discs. However, because chalcogenides are comprised of rare and expensive elements, numerous researchers are exploring new materials that can replace chalcogenides in next-generation optical storage media [5–19].

In our efforts to find a suitable next-generation optical storage medium, we synthesized Ti_3O_5 nanoparticles with a novel phase, λ-Ti_3O_5, in 2010 [20]. Irradiating λ-Ti_3O_5 with laser light at room temperature causes a reversible photo-induced metal-to-semiconductor phase transition between black metallic λ-Ti_3O_5 and brown semiconducting β-Ti_3O_5 (figure 1). λ-Ti_3O_5 is an environmentally friendly and sustainable material that consists of highly abundant elements [21,22]. Consequently, λ-Ti_3O_5 is a promising candidate for use in rewritable recording optical media.

The morphology of λ-Ti_3O_5 depends on its synthesis [20]. A combination of reverse–micelle and sol–gel techniques provides λ-Ti_3O_5 nanocrystals dispersed in a SiO_2 matrix, where the nanocrystal size is 21±11 nm. In contrast, a sintering method, such as sintering titanium dioxide (TiO_2) nanoparticles, provides the flake form λ-Ti_3O_5, where the λ-Ti_3O_5 size is several micrometers assembled from 25±15 nm nanocrystals. The possibility to reducing the size of λ-Ti_3O_5 is important from the viewpoints of basic and applied science. In this work, we demonstrate that tuning the sintering temperature can control

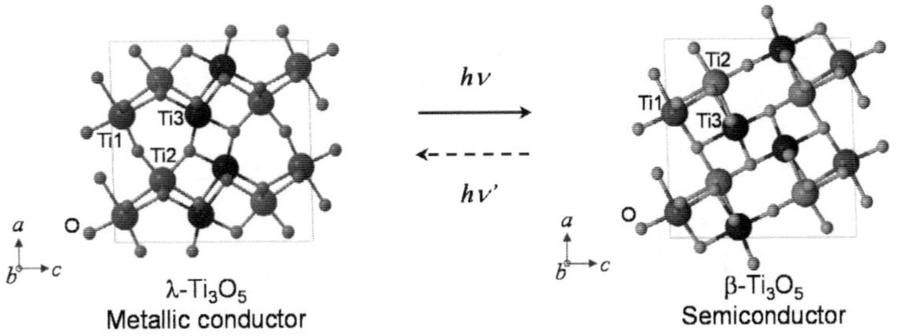

Figure 1. Schematic illustration of the reversible photo-induced metal-to-semiconductor phase transition between λ-Ti_3O_5 and β-Ti_3O_5. Light blue (gray in black and white print), purple (light gray), and dark blue (dark gray) balls represent Ti(1), Ti(2), and Ti(3) atoms of λ-Ti_3O_5, while red (gray), light red (light gray), and dark red (dark gray) balls represent Ti(1), Ti(2), and Ti(3) atoms of β-Ti_3O_5, respectively. Small gray balls represent O atoms.

the size of the λ-Ti_3O_5 nanocrystals prepared via sol–gel synthesis, and report λ-Ti_3O_5 nanocrystals below ten nanometers size.

2. Experimental

2.1. Synthesis

The sol–gel method is employed to synthesize λ-Ti_3O_5 nanocrystals (figure 2). While stirring, 9.6 ml of aqueous ammonia (6.8×10^{-2} mol dm^{-3}) was added to 420 ml of a $TiCl_4$ aqueous solution (2.8×10^{-2} mol dm^{-3}). Next, 24 ml of tetraethoxysilane (TEOS) was added, and the solution was stirred for 20 hours. Then the obtained gel was collected by centrifugation, washed with methanol and chloroform, and dried at 60°C for one day. The resulting dry powder was sintered in hydrogen at a flow rate of 0.7 L min^{-1} for five hours at various temperatures, i.e., 1123°C (sample **1**), 1133°C (sample **2**), 1143°C (sample **3**), 1153°C (sample **4**), 1163°C (sample **5**), 1173°C (sample **6**), 1200°C (sample **7**), and 1250°C (sample **8**). Additional samples were prepared at a fixed sintering temperature of 1163°C for five hours, but with various hydrogen flow rates, i.e., 0.05 L min^{-1} (sample **9**), 0.3 L min^{-1} (sample **10**), 1.5 L min^{-1} (sample **11**), and 3.0 L min^{-1} (sample **12**).

2.2. Characterization

Transmission electron microscope (TEM) measurements were acquired with JEOL JEM-2000EXII. The X-ray diffraction (XRD) measurements were performed using Rigaku Ultima IV with Cu Kα radiation (λ =1.5418 Å). Rietveld analyses were performed using the Rigaku PDXL program.

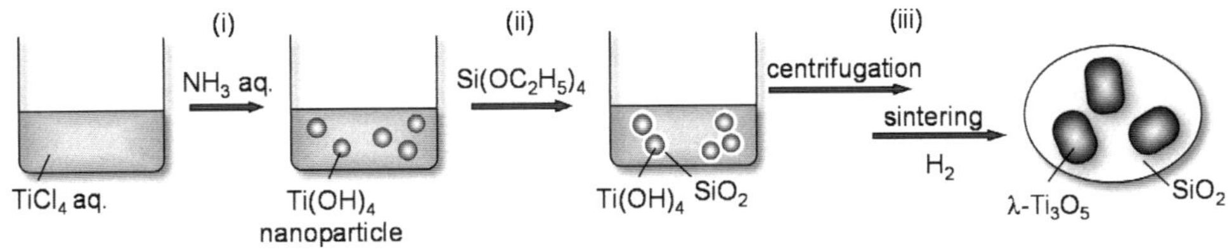

Figure 2. Sol–gel synthesis of λ-Ti_3O_5. (i) Aqueous ammonia is added to an aqueous solution containing titanium chloride, yielding the sol of titanium hydroxide, $Ti(OH)_4$. (ii) TEOS, $Si(OC_2H_5)_4$, is added to the reaction solution to yield a metal hydroxide coated with a sol of silica via hydrolysis, $Ti(OH)_4/SiO_2$. (iii) Centrifuging and sintering the obtained gel under a hydrogen atmosphere yields λ-Ti_3O_5 nanocrystals in a SiO_2 matrix.

3. Results and discussion

3.1. Effect of sintering temperature

Figures 3 show the XRD patterns for samples **1**–**8** at room temperature. The insets show the TEM

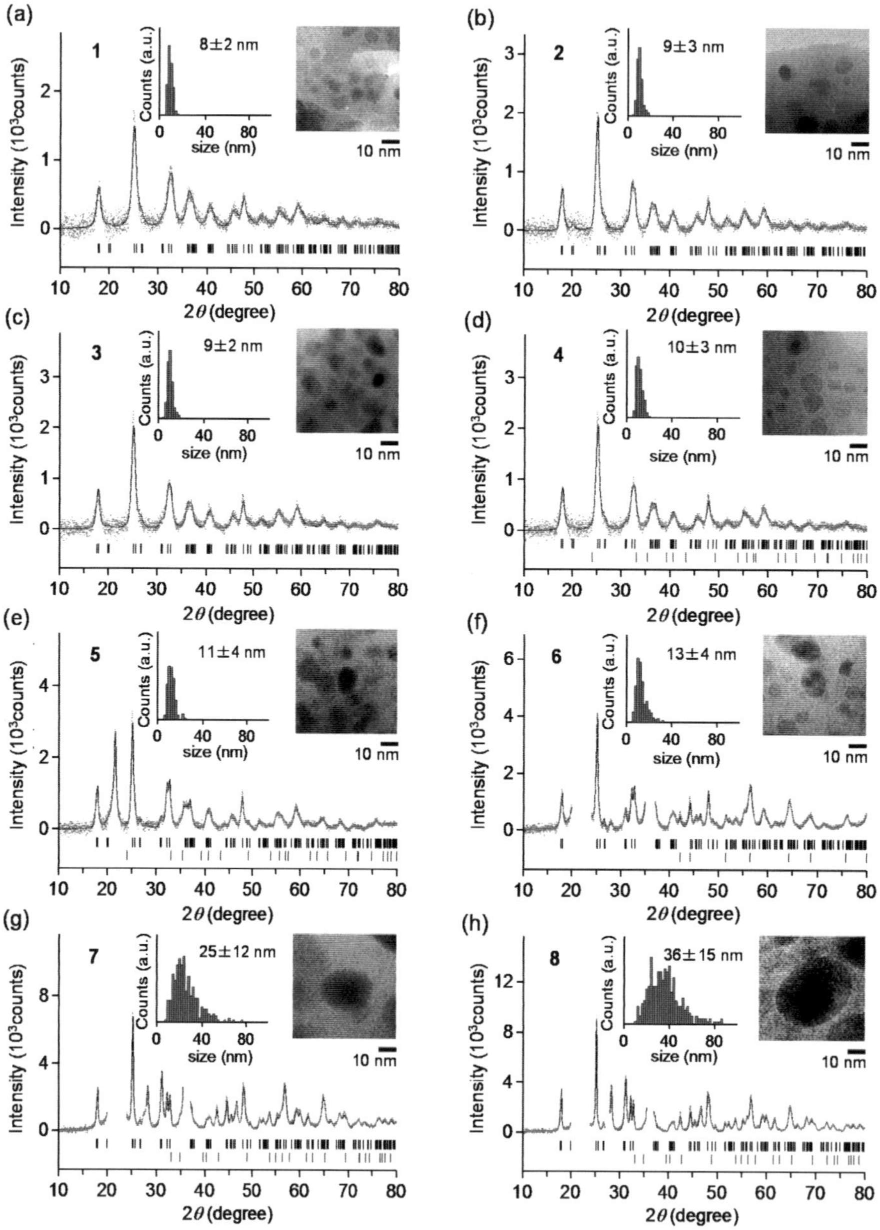

Figure 3. Powder X-ray diffraction patterns and Rietveld analyses for samples sintered at (a) 1123°C (sample **1**), (b) 1133°C (**2**), (c) 1143°C (**3**), (d) 1153°C (**4**), (e) 1163°C (**5**), (f) 1173°C (**6**), (g) 1200°C (**7**), and (h) 1250°C (**8**). Broad baseline caused by amorphous SiO_2 is eliminated. Red (gray in black and white print) dots and black lines are the observed plots and calculated patterns, respectively. Black and gray bars represent the calculated positions of the Bragg reflections of λ-Ti_3O_5 and Ti_2O_3, respectively. Huge peaks of SiO_2 cristobalite are excluded from the Rietveld analyses for samples **6**–**8** (blank ranges). Insets show the TEM images and crystal size distribution.

images and crystal size distribution. Rietveld analyses indicate that the crystal structures of the main phases of samples **1–8** are monoclinic (space group *C*2/*m*). The obtained lattice constants as shown in table 1 are consistent with the previously reported crystal structure parameters for λ-Ti_3O_5 [20]. For samples **1**–**3**, the XRD patterns indicate that the λ-Ti_3O_5 fraction is 100%, whereas those for samples **4**–**8** contain minor peaks, which are assigned to a few percent of Ti_2O_3 impurity.

Figure 4 plots the crystal size versus sintering temperature. Crystals of samples **1**–**8** have average sizes between 8 and 36 nm, i.e., 8±2 nm, 9±3 nm, 9±2 nm, 10±3 nm, 11±4 nm, 13±4 nm, 25±12 nm, and 36±15 nm for samples **1**, **2**, **3**, **4**, **5**, **6**, **7**, and **8**, respectively. A low sintering temperature suppresses crystal growth and the lowest sintering temperature in this study, 1123°C, produced the smallest λ-Ti_3O_5 nanocrystals of 8 nm in size.

Table 1. Crystallographic data of samples **1–8**, including lattice parameters of λ-Ti_3O_5 obtained by Rietveld refinement of the powder XRD patterns.

Sample	**1**	**2**	**3**	**4**
Sintering temperature (°C)	1123	1133	1143	1153
H_2 flow rate (L min^{-1})	0.7	0.7	0.7	0.7
Crystal system	monoclinic	monoclinic	monoclinic	monoclinic
Space group	*C*2/*m*	*C*2/*m*	*C*2/*m*	*C*2/*m*
a (Å)	9.832(4)	9.799(3)	9.832(3)	9.806(3)
b (Å)	3.7962(8)	3.7869(5)	3.7954(6)	3.7931(5)
c (Å)	9.962(4)	9.973(3)	9.984(3)	9.997(2)
β (°)	91.06(3)	91.09(2)	91.02(2)	91.04(2)
V ($Å^3$)	371.8(2)	370.01(17)	372.51(16)	371.78(14)
Z	4	4	4	4
R_{wp} (%)	1.90	1.92	1.88	1.85
S	1.0432	1.0589	1.0458	1.0363
λ-Ti_3O_5 (%)	100	100	100	95.6(4)
Impurity (Ti_2O_3) (%)	0	0	0	4.4(4)
Sample	**5**	**6**	**7**	**8**
Sintering temperature (°C)	1163	1173	1200	1250
H_2 flow rate (L min^{-1})	0.7	0.7	0.7	0.7
Crystal system	monoclinic	monoclinic	monoclinic	monoclinic
Space group	*C*2/*m*	*C*2/*m*	*C*2/*m*	*C*2/*m*
a (Å)	9.8327(18)	9.8158(16)	9.8244(12)	9.8202(10)
b (Å)	3.7952(3)	3.7898(3)	3.7829(2)	3.7818(3)
c (Å)	9.9794(15)	9.9725(13)	9.9545(11)	9.9567(8)
β (°)	91.008(11)	91.038(9)	91.041(8)	91.081(5)
V ($Å^3$)	372.34(9)	370.92(8)	369.90(6)	369.71(4)
Z	4	4	4	4
R_{wp} (%)	2.09	2.58	3.69	3.58
S	1.1845	1.327	1.8241	1.7675
λ-Ti_3O_5 (%)	95.2(4)	91.4(4)	94.4(5)	97.0(6)
Impurity (Ti_2O_3) (%)	4.8(4)	8.6(3)	5.6(3)	3.0(3)

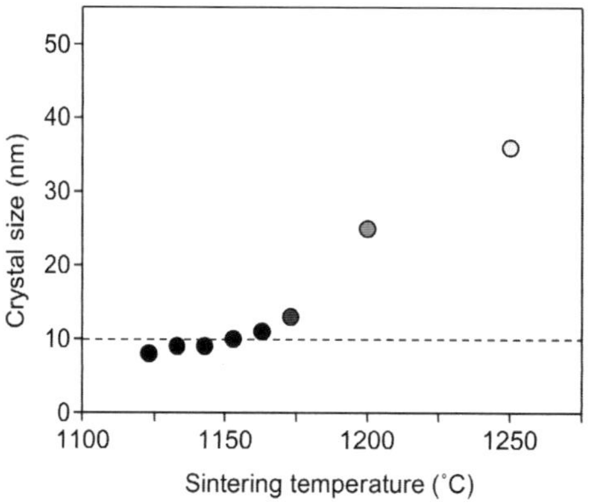

Figure 4. Crystal size versus sintering temperature under a hydrogen flow rate of 0.7 L min^{-1}.

3.2. Effect of hydrogen flow rate

Samples **9–12** were sintered at a fixed temperature of 1163°C, but the hydrogen flow rate was changed between 0.05–3.0 L min^{-1}. Figures 5 show the XRD patterns of samples **9**–**12**. The insets are

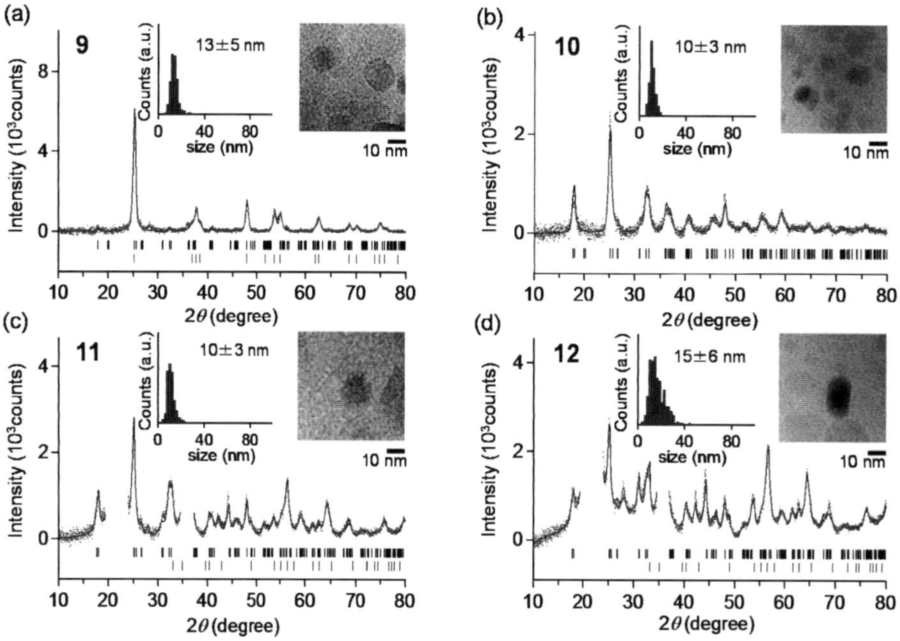

Figure 5. Powder X-ray diffraction patterns and Rietveld analyses for samples sintered at a fixed temperature of 1163°C under hydrogen flow rate of (a) 0.05 L min^{-1} (sample **9**), (b) 0.3 L min^{-1} (**10**), (c) 1.5 L min^{-1} (**11**), and (d) 3.0 L min^{-1} (**12**). Broad baseline caused by amorphous SiO_2 is eliminated. Blue (gray in black and white print) dots and black lines are the observed plots and calculated patterns, respectively. Black and gray bars represent the calculated positions of the Bragg reflections of λ-Ti_3O_5 and TiO_2 or Ti_2O_3, respectively. Huge peaks of SiO_2 cristobalite are excluded from the Rietveld analyses for samples **11** and **12** (blank ranges). Insets show the TEM images and crystal size distribution.

TEM images and size distribution. Rietveld analyses indicate that the main phase of sample **9** is anatase TiO_2, while samples **10–12** are mainly comprised of λ-Ti_3O_5 (table 2). Samples **11** and **12** contain a minor impurity phase of Ti_2O_3, and the Ti_2O_3 fraction increases as the hydrogen flow rate increases. Samples **9–12** have an average size of ten or more nanometers, i.e., varying the hydrogen flow rate has less effect on the nanocrystal size, compared to varying the sintering temperature.

Table 2. Crystallographic data of samples **9–12**, including lattice parameters of λ-Ti_3O_5 obtained by Rietveld refinement of the powder XRD patterns.

Sample	**9**	**10**	**11**	**12**
Sintering temperature (°C)	1163	1163	1163	1163
H_2 flow rate (L min^{-1})	0.05	0.3	1.5	3.0
Crystal system	monoclinic	monoclinic	monoclinic	monoclinic
Space group	*C*2/*m*	*C*2/*m*	*C*2/*m*	*C*2/*m*
a (Å)	9.883(17)	9.820(2)	9.791(3)	9.794(5)
b (Å)	3.791(5)	3.7912(5)	3.7874(5)	3.7767(9)
c (Å)	9.924(16)	9.980(2)	9.962(2)	9.949(4)
β (°)	90.72(12)	91.068(18)	90.895(18)	90.67(4)
V ($Å^3$)	369.4(10)	371.49(12)	369.37(14)	368.0(3)
Z	4	4	4	4
R_{wp} (%)	2.22	2.16	2.49	3.47
S	1.2584	1.1775	1.2775	1.7190
λ-Ti_3O_5 (%)	10.4(5)	100	79.5(5)	62.9(11)
Anatase TiO_2 (%)	89.6(6)	0	0	0
Impurity (Ti_2O_3) (%)	0	0	20.5(3)	37.1(9)

3.3. Temperature dependence of crystal structure

Variable-temperature XRD measurements were conducted for the λ-Ti_3O_5 nanocrystals. Figure 6 plots the peak position versus temperature for the XRD patterns of sample **10** measured in the range of 31.4–

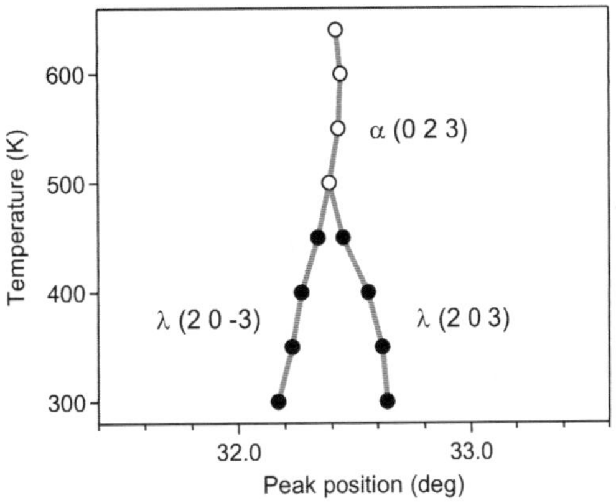

Figure 6. Peak position versus temperature of the XRD patterns in the angle range of 31.4–33.5° for sample **10**. Diffraction peaks of λ-Ti_3O_5 (2 0 −3) and (2 0 3) (filled circles) change into the α-Ti_3O_5 (0 2 3) peak (open circles) as the temperature increases.

33.5°. Upon heating, the diffraction peaks of λ-Ti_3O_5 transform into the α-Ti_3O_5 peak with an orthorhombic crystal structure (*Cmcm*), e.g., (2 0 −3) and (2 0 3) λ-Ti_3O_5 peaks transform into the (0 2 3) α-Ti_3O_5 peak around 500 K. Cooling to 300 K restores λ-Ti_3O_5. The observed phase transition temperature of 500 K is consistent with the reported second-order phase transition temperature between λ-Ti_3O_5 and α-Ti_3O_5 for nanocrystals obtained by a combination of reverse–micelle and sol–gel techniques [20].

3.4. Mechanism

Next, we considered the reason why λ-Ti_3O_5 forms as nanocrystals by this sol–gel synthesis. In this method, the precursor $Ti(OH)_4$ particles are uniformly dispersed in the aqueous solution. The hydrolysis of TEOS into SiO_2 occurs slowly and homogeneously in the sol–gel process, resulting in a homogeneous SiO_2 shell over the precursor particle surfaces. Due to the SiO_2 matrix, the sintering process suppresses crystalline growth. Consequently, a combination of the sol–gel method and sintering process is suitable to form λ-Ti_3O_5 in the size of ca. ten nanometers. Furthermore, the dependence of the crystal size on temperature is due to the viscosity of the SiO_2 matrix; the viscosity of SiO_2 increases as the temperature decreases, which results in smaller nanocrystals at lower temperature.

4. Conclusion

Synthesizing small-sized nanocrystals of a photo-induced phase transition material is a promising method to develop high-density recording media. Herein we demonstrate that tuning the sintering temperature can control the size of the λ-Ti_3O_5 nanocrystals. A sintering temperature of 1123°C gives the smallest λ-Ti_3O_5 nanocrystals, 8 nm. Adjusting the hydrogen flow rate can tune the oxidation-reduction state of the sample without apparent change in the crystal size. The memory density in the recording media is estimated to be 1 terabit inch^{-2} with nanocrystal size of 21 nm, which is our previously reported crystal size [20], while the density is expected to achieve 10 terabit inch^{-2} with 8 nm nanocrystals. Although we prepared the smallest sized λ-Ti_3O_5 nanocrystals in this work, a study to synthesize even smaller nanocrystals is currently underway.

Acknowledgements

We are grateful to Y. Kakegawa, H. Tsunakawa, and M. Adachi for acquiring the TEM images, and M. Yoshikiyo for useful discussion. The present research was supported partly by the Core Research for Evolutional Science and Technology (CREST) program of the Japan Science and Technology Agency (JST), Next Generation World-Leading Researchers (NEXT) project from the Japan Society for the Promotion of Science (JSPS), a Grant for the Global COE Program "Chemistry Innovation through Cooperation of Science and Engineering", Advanced Leading Graduate Course for Photon Science (ALPS) from the Ministry of Education, Culture, Sports, Science and Technology of Japan (MEXT), and the Asahi Glass Foundation. T. N. was grateful to the JSPS through Program for Leading Graduate Schools (MERIT). We also recognize the Cryogenic Research Center, The University of Tokyo, and the Center for Nano Lithography & Analysis, The University of Tokyo, which are supported by MEXT.

References

[1] Yamada N, Ohno E, Nishiuchi K, Akahira N and Takao M 1991 *J. Appl. Phys.* **69** 2849–56
[2] Kolobov A V, Fons P, Frenkel A I, Ankudinov A L, Tominaga J and Uruga T 2004 *Nature Mater.* **3** 703–708
[3] Wuttig M and Yamada N 2007 *Nature Mater.* **6** 824–832
[4] Sun Z, Zhou J and Ahuja R 2006 *Phys. Rev. Lett.* **96** 055507
[5] Nasu K 1997 *Relaxations of Excited Stats and Photo-Induced Structural Phase Transition* (Japan: Springer)
[6] Gütlich P, Hauser A and Spiering H 1994 *Angew. Chem. Int. Ed. Engl.* **33** 2024–54
[7] Kahn O and Martinez C J 1998 *Science* **279** 44–48
[8] Decurtins S, Gütlich P, Köhler C P, Spiering H and Hauser A 1984 *Chem. Phys. Lett.* **105** 1–4
[9] Koshihara S, Tokura Y, Mitani T, Saito G and Koda T 1990 *Phys. Rev. B* **42** 6853–56

[10] Verdaguer M 1996 *Science* **272** 698–699
[11] Herrera J M, Marvaud V, Verdaguer M, Marrot J, Kalisz M and Mathonière C 2004 *Angew. Chem. Int. Ed.* **43** 5468–71
[12] Dei A 2005 *Angew. Chem. Int. Ed.* **44** 1160–63
[13] Tokoro H, Ohkoshi S and Hashimoto K 2003 *Appl. Phys. Lett.* **82** 1245–47
[14] Ohkoshi S, Hamada Y, Matsuda T, Tsunobuchi Y and Tokoro H 2008 *Chem. Mater.* **20** 3048–54
[15] Tokoro H and Ohkoshi S 2011 *Dalton Trans.* **40** 6825–33
[16] Ohkoshi S and Tokoro H 2012 *Acc. Chem. Res.* **45** 1749–58
[17] Ohkoshi S, Imoto K, Tsunobuchi Y, Takano S and Tokoro H 2011 *Nature Chem.* **3** 564–569
[18] Miyano K, Tanaka T, Tomioka Y and Tokura Y 1997 *Phys. Rev. Lett.* **78** 4257–60
[19] Fiebig M, Miyano K, Tomioka Y and Tokura Y 1998 *Science* **280** 1925–28
[20] Ohkoshi S, Tsunobuchi Y, Matsuda T, Hashimoto K, Namai A, Hakoe F and Tokoro H 2010 *Nature Chem.* **2** 539–545
[21] Prassides K 2010 *Nature Chem.***2** 517–519
[22] Nakamura E and Sato K 2011 *Nature Mater.* **10** 158–161

Toward Si-based high-efficiency thin-film solar cells using semiconducting $BaSi_2$

Takashi Suemasu

Institute of Applied Physics, University of Tsukuba, 1-1-1 Tennohdai, Tsukuba, Ibaraki 305-8573, Japan

Email: suemasu@bk.tsukuba.ac.jp

Abstract. We have grown $BaSi_2$ epitaxial films on Si(111) substrates by molecular beam epitaxy, and investigated their optical properties such as optical absorption coefficients, minority-carrier diffusion length and minority-carrier lifetime. These are key parameters which determine the performance of solar cells. The band gap of $BaSi_2$ was measured to be approximately 1.3 eV. The absorption coefficient reached approximately 3×10^4 cm^{-1} at 1.5 eV. The minority-carrier diffusion length and minority-carrier lifetime were found to be about 10 μm and 8 μs, respectively. These values are great enough for thin-film solar cell applications. Control of carrier type and carrier concentration was also demonstrated.

1. Introduction

Thin-film solar cell materials such as CIGS, CZTS, CdTe, and organic materials have been attracting increasing attention due to their high efficiency and low cost. In contrast, the optical absorption layers of crystalline Si solar cells tend to be much thicker than conventional thin-film solar cells, because the optical absorption coefficient α is much smaller for crystalline Si. Therefore, novel Si-based materials for high-efficiency thin-film solar cells have received significant interest. However, little steadfast effort has been devoted to any materials other than Si, CZTS, CIGS, CdTe and III-V compounds as far as inorganic semiconductors are concerned. Among such materials, we have focused much attention on semiconducting $BaSi_2$. The band gap of $BaSi_2$ is approximately 1.3 eV,[1,2] and can be increased up to 1.4 eV in $Ba_{1-x}Sr_xSi_2$,[3] which matches the ideal solar spectrum much better than crystalline Si. In addition, $BaSi_2$ has a very large absorption coefficient of approximately 3×10^4 cm^{-1} at 1.5 eV.[2] A large value of α and expansion of the band gap in $Ba_{1-x}Sr_xSi_2$ were theoretically expected.[4,5] In solar cells, both absorption coefficient α and minority-carrier diffusion length L are critical. In order to efficiently extract photogenerated carriers, the diffusion length should be larger than the absorption length. Direct band gap semiconductors like GaAs usually have large absorption coefficient. But this generally comes with faster recombination and hence relatively smaller diffusion length. In contrast, both α and L were found to be large enough in $BaSi_2$ (L~10μm).[2,6] This is because $BaSi_2$ is an indirect band gap semiconductor with a direct transition 0.1 eV above the band edge. Recently, we successfully achieved large photoresponsivity and internal quantum efficiency exceeding 70% in 0.4μm-thick *a*-axis-oriented $BaSi_2$ epitaxial layers.[7] Excess carrier recombination mechanisms in $BaSi_2$ have also been studied.[8] These results have spurred interest in this material.

Control of the conductivity of $BaSi_2$ by impurity doping is a requirement because the basic structure of a solar cell is a *p-n* junction. According to Imai and Watanabe,[9] substitution of Si in the $BaSi_2$ lattice is more favorable than substitution of Ba from an energetic point of view by first-principles calculation. In our previous works, the electron concentration of Sb-doped n-type $BaSi_2$ was controlled in the range between 10^{16} and 10^{20} cm^{-3} at room temperature (RT).[10] Very recently, we have achieved the hole concentration exceeding 10^{19} cm^{-3} in B-doped p-type $BaSi_2$. The hole concentration of B-doped $BaSi_2$ was controlled in the range between 10^{16} and 10^{20} cm^{-3} at RT. In this article, the current status and future prospect of this material towards high-efficiency and thin-film solar cell applications are reviewed.

2. Experimental procedures

An ultrahigh vacuum (UHV) chamber equipped with a Knudsen cell for Ba and an electron beam gun for Si was employed. Before the growth, the Si(111) substrates were prepared by subjecting them to the following treatment. The substrates were washed using RCA clean steps, which removed organic and metallic contaminants. The substrates were then annealed at 830 °C for 30 min in the UHV (1×10^{-6} Pa) chamber to remove the protective SiO_2 layers. After annealing, a 7×7 streaky reflection high-energy electron diffraction (RHEED) pattern was observed, indicating a clean Si surface. A two-step growth method was adopted, which included reactive deposition epitaxy (RDE; Ba deposition on hot Si) and molecular beam epitaxy (MBE; co-deposition of Ba and Si) to form thick $BaSi_2$ films.[11,12] The RDE process was carried out for deposition of a template layer as a $BaSi_2$ precursor prior to the subsequent MBE process. The same growth method was successfully utilized for the epitaxial growth of semiconducting β-$FeSi_2$ films on both Si(001) and Si(111) substrates.[13,14] The crystalline quality of the films was evaluated using RHEED and θ-2θ X-ray diffraction (XRD) measurements using Kα x-rays.

3. Results and discussions

3.1 Optical absorption properties

For measurements of optical absorption coefficients, we fabricated the silicon-on-insulator (SOI) substrates using wafer bonding at RT as well as chemical mechanical polishing (CMP). First, 500-μm-thick high-resistive FZ-Si(111) ($\rho > 1000$ Ω·cm) and 500-μm-thick fused silica wafers were bonded at RT. Then, the Si wafer was mechanically ground and polished by CMP down to about 0.7 μm thickness. Prior to loading the substrates into the MBE chamber, oxide layers on the surface of the SOI substrate were etched away by diluted hydrofluoric acid. Thermal treatment was performed at 590 °C for 20 min in UHV to clean the hydrogen-terminated surface. A 20-nm-thick $BaSi_2$ template layer was then formed at 525 °C using RDE, prior to the deposition of an 80-nm-thick $BaSi_2$ layer by MBE (sample A). Si and Ba atoms were co-deposited at 575°C by an electron-beam gun and a Knudsen cell, respectively. For comparison, $BaSi_2$ layers were also epitaxially grown on a Si(111) substrate (sample B) in the same manner as described previously.[12] RHEED and θ-2θ XRD patterns were utilized to evaluate the crystalline quality of the $BaSi_2$ layers. We employed a JASCO U-best 570 spectrophotometer to obtain the transmission spectra of the samples. We observed the (1×1) streaky pattern of Si(111) after thermal cleaning at 590°C.

Figures 1(a) and 1(b) show RHEED patterns of the RDE-grown $BaSi_2$ template layer. The incident electron beam was along the (a) Si[1-10] and (b) Si[11-2] directions. After the RDE growth, we confirmed that the $BaSi_2$ film was epitaxially grown on the Si layer of the SOI substrate as shown in Figs. 1(c) and 1(d). Since these patterns could also be observed at the end of MBE growth, we determined that the $BaSi_2$ film was indeed grown epitaxially.

Figure 2 shows the θ-2θ XRD patterns of samples A and B, $BaSi_2$ films grown on the SOI and Si(111) substrates, respectively. We can see diffraction peaks only from (100)-oriented $BaSi_2$ planes, such as the (200), (400), (600) planes. For the $BaSi_2$ epitaxial film on the SOI substrate, the peak intensities of these planes were almost the same as those in the $BaSi_2$ epitaxial film on the Si(111) substrate. These results indicate that the crystalline quality of the $BaSi_2$ film on the SOI substrate was equivalent to that on the single-crystalline Si(111) substrate.

The transmission spectrum for the $BaSi_2$/SOI structure is presented in Fig. 3. This spectrum was significantly influenced by an interference effect within the 0.7-μm-thick Si layers. The Si layers were very thin and flat due to the CMP process, so that these interference fringes were superimposed on the spectrum. We used the equations presented in Ref. 15 to derive the interference-free transmission spectrum. As shown in Fig. 3, the transmission spectrum was fitted by the maximal extremes of the interference fringes (T_M) and also by their minimal extremes (T_m). Assuming that the Si layers have a certain absorption coefficient on a transparent substrate, the interference-free transmission spectrum T_α can be expressed as just the geometric mean of T_M and T_m ($T_\alpha=\sqrt{T_M T_m}$) over the entire region of the transmission spectrum. We calculated the T_α values using the T_M and T_m curves, and also eliminated the absorption due to the Si layers by measuring the transmission spectrum of another SOI substrate. We finally normalized T_α by taking the reflectivity of the sample into consideration. Using T_α, we could obtain the absorption spectrum shown in Fig. 4(a). The absorption coefficient of $BaSi_2$ reached 3×10^4 cm^{-1} at 1.5 eV. Figure 4(b) shows the $(\alpha dh\nu)^{1/2}$ versus $h\nu$ plot for deriving the indirect optical absorption edge. The straight fit line intersects the x-axis at 1.34 eV. Thus, the indirect absorption edge with phonon emission was 1.34 eV. The absorption with phonon absorption is neglected, because the fitting was performed at the higher-energy region than the indirect band gap. Thus, the indirect band gap is determined to be approximately 1.32 eV, assuming that the phonon energy is approximately 25 meV. It is true that absorption coefficients below 10^2 cm^{-1} are necessary to determine the precise energy gap. Much thicker $BaSi_2$ enables us to measure smaller absorption coefficients free from errors. In addition, we can't rule out the possibility that defects contribute to absorption to some extent in the energy range below 1.34 eV.

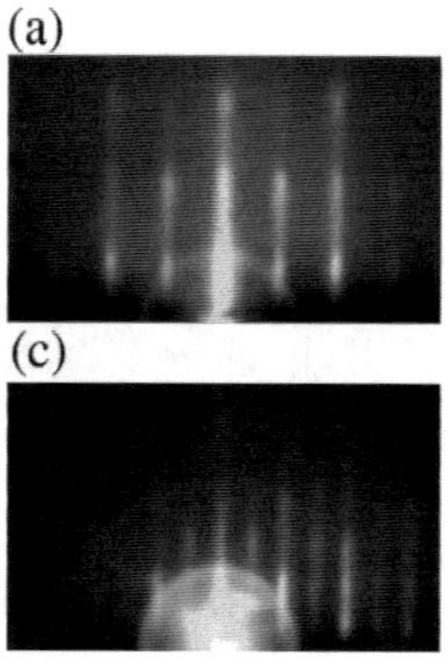

Figure 1. RHEED patterns of $BaSi_2$ in sample A, observed after (a)(b) RDE and (c)(d)MBE growth along the (a)(c) Si[1-10] and (b)(d) Si[11-2] azimuths.

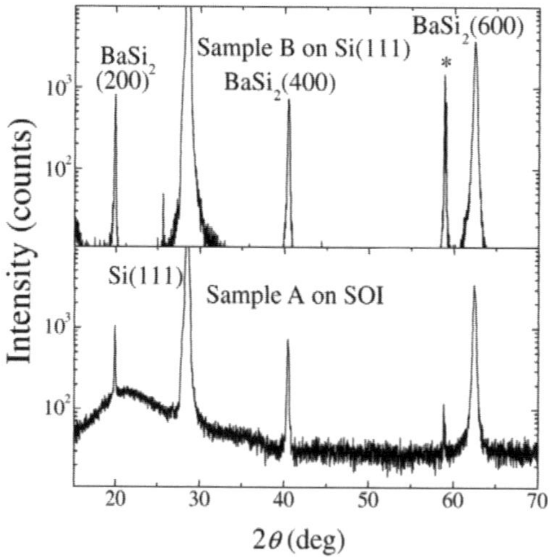

Figure 2. θ-2θ XRD patterns for samples A and B, grown on SOI and Si(111) substrates, respectively. The forbidden diffraction peak of Si(222) indicated by asterisk occurs by double diffraction.

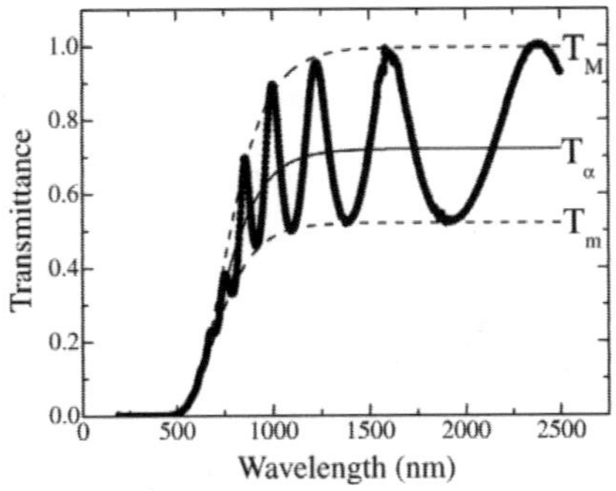

Figure 3. Transmission spectrum for sample A, where $T_\alpha = \sqrt{T_M T_m}$.

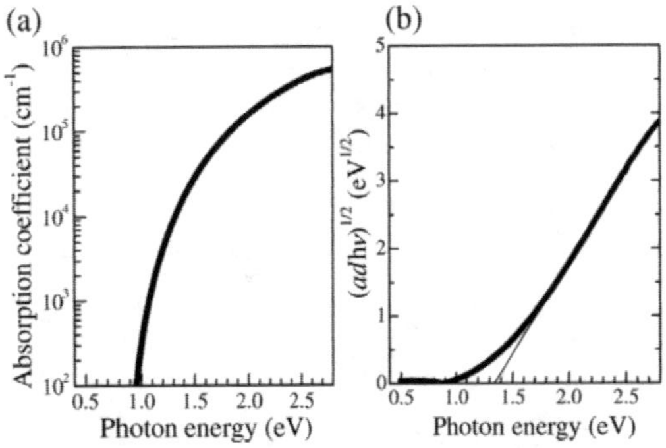

Figure 4. (a) Absorption spectrum for sample A. (b) $(\alpha d h\nu)^{1/2}$ versus $h\nu$ plot for sample A.

3.2 Minority-carrier properties

3.2.1. Minority-carrier diffusion length. For measurements of minority-carrier diffusion length in undoped $BaSi_2$, we fabricated 300-nm-thick undoped n-$BaSi_2$ films by RDE at 550 °C for 5 min, followed by MBE at 600 °C for 120 min. Undoped $BaSi_2$ shows n-type conductivity with electron concentrations of approximately 10^{16} cm^{-3}.[1] In order to investigate the grain size of $BaSi_2$ and grain boundaries (GBs), plan-view transmission electron microscopy (TEM) samples prepared by mechanical polishing and ion milling were observed using TOPCON EM-002B operated at 120 kV. For electron beam induced current (EBIC) measurements, Al/n-$BaSi_2$ Schottky diodes were formed. Front-side Schottky contacts were formed with Al on the $BaSi_2$ surface via wire bonding, and the back-side ohmic contact was made with Al by sputtering. EBIC observations were carried out in the edge-scan configuration with a Hitachi S4300 field-emission scanning electron microscope (SEM) in the EBIC mode at RT. The acceleration voltage of the electron beam, V_{ac}, was set at 5 kV to avoid penetration of the beam into the Si substrate. The penetration depth of the electron beam is estimated to be shorter than 300 nm, which is the thickness of the $BaSi_2$ layers, when V_{ac} is 5 kV, with the density of $BaSi_2$ being 5.14 g/cm^3.

Figures 5(a)(b) and 5(c)(d) show secondary-electron (SE) and EBIC images around the Al contact, respectively, with V_{ac} = 5 kV. In the EBIC method, carriers generated within the diffusion length in the n-type $BaSi_2$ are collected by the electric field under the Al contact and sensed as a current in the external circuit. In Figs. 5(c)(d), the brighter regions show higher collection of electron-beam-induced carriers in the $BaSi_2$. We cannot see defect-related black lines. Figure 6 shows the EBIC line-scan data along dotted line AA' in Fig. 5(c). The EBIC profile shows an exponential dependence of the distance from the Al contact. In this work, the diffusion length of minority carriers was roughly estimated to be approximately 10 μm, assuming that the EBIC profile varies as $\exp(-x/L)$, where x is the distance from the Al edge (point A) along the dotted line, and L is the diffusion length of holes for $BaSi_2$. The obtained minority-carrier diffusion length is much larger than the grain size of the $BaSi_2$, implying that the GBs do not work as defect centers for minority carriers in n-$BaSi_2$. The contribution of carriers generated within the n-Si substrate to the measured EBIC signals can be excluded, because the simulated penetration depth of the electron beam is shorter than 300 nm, the thickness of the $BaSi_2$ layers, when V_{ac} is 5 kV. Thus, it is reasonable to think that the number of carriers generated within the Si substrate was negligibly small compared to those generated in the $BaSi_2$ layers. To confirm the GBs character, an EBIC system with spatial resolutions less than 0.1 μm may be necessary. We should also note here that the obtained minority-carrier diffusion length is roughly 30 times larger than $1/\alpha$ (=0.3 μm) at 1.5 eV, suggesting that this value is large enough for solar cell applications.

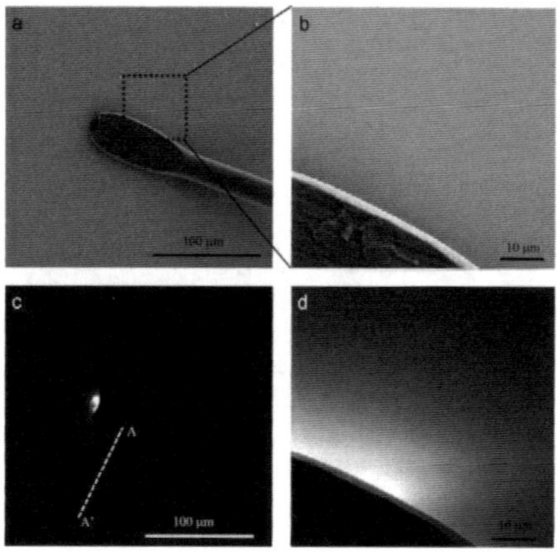

Figure 5. (a)(b) SE and (b)(d) EBIC images around the Al contact.

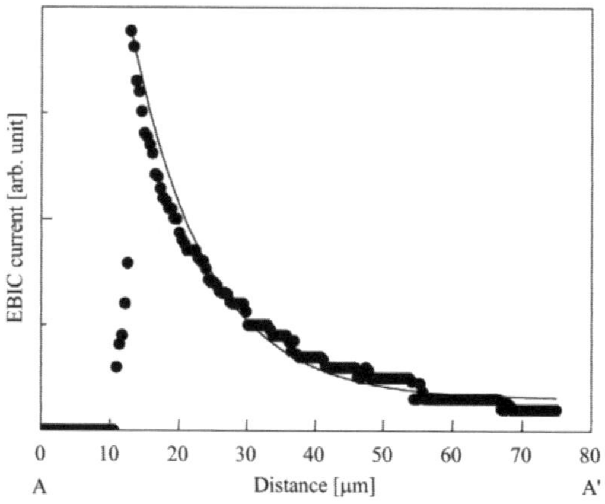

Figure 6. Experimental and simulated (solid line) EBIC line-scan profiles along the dotted line from points A to A' in Fig. 5(c).

3.2.2 Minority-carrier lifetime diffusion length. After cleaning a floating-zone n-Si(111) substrate (ρ>1000 Ω·cm) at 900 °C for 30 min in UHV, a 5-nm-thick $BaSi_2$ template layer was grown on the Si substrate by RDE, followed by MBE to form a 1.5-μm-thick $BaSi_2$ epitaxial film. Figure 7 shows the relationship between apparent minority-carrier life time τ_a and excess-carrier density of the 1.5-μm-thick $BaSi_2$. The undoped $BaSi_2$ shows n-type conductivity and the electron concentration is about 10^{16} cm^{-3}.[1] The excess carrier concentration was calculated from the absorption coefficient of $BaSi_2$ at 349 nm, and the irradiated laser intensity, which was varied from 1.1×10^2 to 1.3×10^2 W/cm^2. The inset shows the decay curve when the excess carrier concentration was 7.4×10^{16} cm^{-3}. Decay can be divided into two modes in terms of decay rate. We see initial rapid decay, followed by approximately constant decay. According to our previous work, the rapid decay is caused by Auger recombination.[8] τ_a was calculated by approximating the tail region of the decay curve by a single exponential curve. We see that τ_a decreases with increasing the excess-carrier density. This is because the excess-carrier density is several orders of magnitudes higher than majority-carrier density at equilibrium in undoped n-$BaSi_2$. Such high carrier injection leads to multicarrier recombination. Therefore, when we discuss about the minority-carrier lifetime, we should use lower injection value. As can be seen in Fig. 7, τ_a becomes independent of excess-carrier density when the excess carrier concentration becomes smaller than 10^{16} cm^{-3}, indicating that τ_a corresponds to the effective minority-carrier lifetime τ in this region. This result makes sense because the electron concentration of undoped n-$BaSi_2$ is about 10^{16} cm^{-3}.[1] Therefore, it is reasonable to think that the minority-carrier lifetime in the 1.5-μm-thick $BaSi_2$ is about 8 μs. Assuming

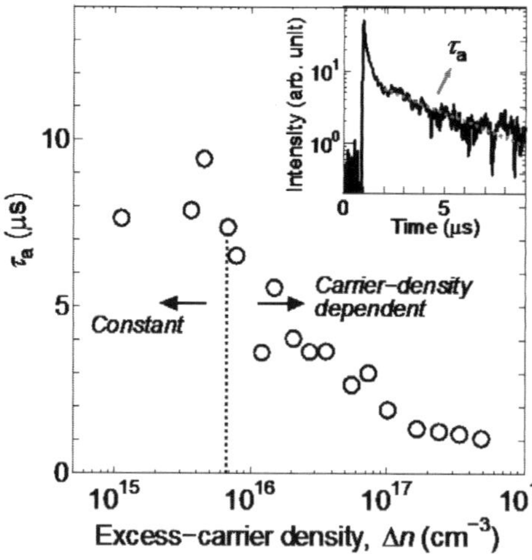

Figure 7. Relationship between minority-carrier life time and excess-carrier density of 1.5-μm-thick $BaSi_2$. The inset shows the photoconductivity decay curve of the 1.5-μm-thick $BaSi_2$ film when the excess-carrier concentration is 7.4×10^{16} cm^{-3}.

that the hole mobility is 1 $cm^2V^{-1}s^{-1}$,[16] the minority-carrier diffusion length L is estimated to be about 4.5 μm from the equation of $L = \sqrt{k_B T \mu\tau / q}$. Here, k_B is the Boltzmann constant, T the absolute temperature, μ the mobility of holes, τ the effective minority-carrier lifetime, and q the elemental charge. This value is about three times larger than the absorption layer thickness (~1.5 μm) in a $BaSi_2$ pn junction diode. On the basis of these results, we achieved the formation of high-quality $BaSi_2$ films thicker than 1.5 μm, where the minority-carrier lifetime was large enough for thin-film solar cell applications.

3.3 Electrical properties of impurity-doped $BaSi_2$ films

The basic structure of a solar cell is a *pn* junction. Thus, the control of carrier type and carrier concentration is a requirement. According to Imai and Watanabe, substitution of Si in the $BaSi_2$ lattice is more favorable than substitution of Ba from the energetic point of view.[5] In line with this theoretical expectation, Sb-doped $BaSi_2$ exhibits *n*-type conductivity, while In-, Al-, and Ag-doped $BaSi_2$ exhibit *p*-type conductivity.[10,16,17] Briefly, a 10-nm-thick $BaSi_2$ epitaxial film was first grown by RDE on high-resistive FZ *n*- and *p*-type Si(111) substrates (ρ>1000 Ω·cm). Then, Sb (B) was coevaporated with Si and Ba to form Sb(B)-doped $BaSi_2$ films by MBE. The electrical properties were characterized by Hall measurements using the van der Pauw method at RT. The applied magnetic field was 0.7 T, normal to the sample surface.

As shown in Fig. 8, as the electron hole concentration increased, the mobility decreased. This trend is usually predicted by ionized impurity scattering in conventional semiconductors. As can be seen, the electron concentration was controlled in the range between 10^{17} and 10^{20} cm^{-3} at RT by changing the temperature of the Sb Knudsen cell crucible. The hole concentration in the B-doped $BaSi_2$ was also controlled in the range between 10^{17} and 10^{20} cm^{-3}. Therefore, the only remaining process was the formation of *p*-type $BaSi_2$ on the *n*-$BaSi_2$ to complete a *p-n* junction diode.

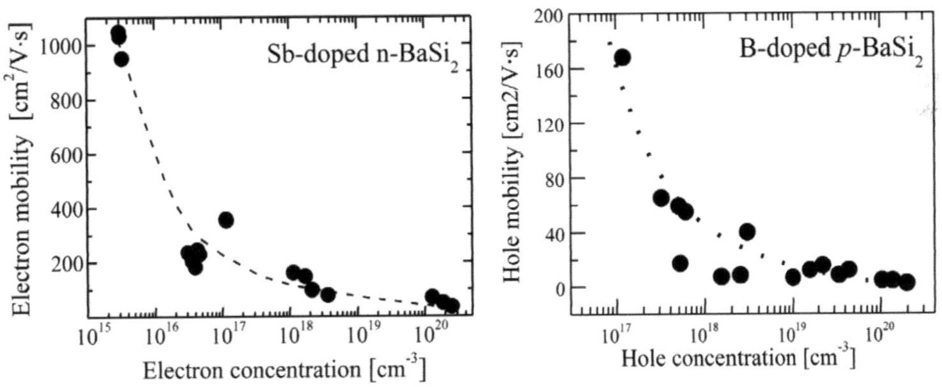

Figure 8. Relationship between electron concentration versus electron mobility in Sb-doped n-type $BaSi_2$ (left) and hole concentration versus hole mobility in B-doped p-type $BaSi_2$ (right). The measurements were performed at RT.

4. Summary

In this article, current status toward Si-based high-efficiency thin-film solar cells was reviewed in terms of optical absorption properties and minority-carrier properties such as minority-carrier diffusion length and minority-carrier lifetime. These are key parameters which determine the performance of solar cells. The band gap of $BaSi_2$ was measured to be approximately 1.3 eV. The absorption coefficient reached approximately 3×10^4 cm^{-1} at 1.5 eV. The minority-carrier diffusion length and minority-carrier lifetime were found to be about 10 μm and 8 μs, respectively. In particular, what's interesting with $BaSi_2$ is that it has large absorption coefficients and at the same time large minority-carrier diffusion length. This feature, caused by the fact that the direct transition starts just 0.1 eV above the band edge, is favourable to effectively extract photogenerated carriers.

References

[1] K. Morita, Y. Inomata, T. Suemasu, Thin Solid Films **508**, 363 (2006).
[2] K. Toh, T. Saito, T. Suemasu, Jpn. J. Appl. Phys. **50**, 068001 (2011).
[3] K. Morita, M. Kobayashi, T. Suemasu, Jpn. J. Appl. Phy. **45**, L390 (2006).
[4] D. B. Migas, V. L. Shaposhnikov, V.E. Borisenko, Phys. Status Solidi B **244**, 2611 (2007).
[5] Y. Imai, and A. Watanabe, Thin Solid Films **515**, 8219 (2007).
[6] M. Baba, K. Toh, K. Toko, N. Saito, N. Yoshizawa, K. Jiptner, T. Sekiguchi, K. O. Hara, N. Usami, T. Suemasu, J. Cryst. Growth **348**, 75 (2012).
[7] W. Du, M. Suzuno, M. A Khan, K. Toh, M. Baba, K. Nakamura, K. Toko, N. Usami, T. Suemasu, Appl. Phys. Lett. **100**, 152114 (2012).
[8] K. O. Hara, N. Usami, K. Toh, M. Baba, K. Toko, and T. Suemasu, J. Appl. Phys. **112**, 083108 (2012).
[9] Y. Imai, A. Watanabe, Intermetallics **15**, 1291 (2007).
[10] M. Kobayashi, Y. Matsumoto, Y. Ichikawa, D. Tsukada, T. Suemasu, Appl. Phys. Express **1**, 05403 (2008).
[11] Y.Inomata, T.Nakamura, T.Suemasu and F.Hasegawa, Jpn. J. Appl. Phys. **43**, 4155 (2004).
[12] Y.Inomata, T.Nakamura, T.Suemasu and F.Hasegawa, Jpn. J. Appl. Phys. **43**, L478 (2004).
[13] M.Tanaka, Y.Kumagai, T.Suemasu and F.Hasegawa, Jpn. J. Appl. Phys. **36**, 3620 (1997).
[14] N.Hiroi, T.Suemasu, K.Takakura, N.Seki and F.Hasegawa, Jpn. J. Appl. Phys. **40**, L1008 (2001).
[15] R. Swanepoel, J. Phys. E **16**, 1214 (1983).
[16] M. Ajmal Khan, K. O. Hara, W. Du, M. Baba, K. Nakamura, M. Suzuno, K. Toko, N. Usami, and T. Suemasu, Appl. Phys. Lett. **102**, 112107 (2013).
[17] M. Ajmal Khan, T. Saito, K. Nakamura, M. Baba, W. Du, K. Toh, K. Toko, and T. Suemasu, Thin Solid Films **522**, 95 (2012).

IOP Conf. Series: Materials Science and Engineering **54** (2014) 012010 doi:10.1088/1757-899X/54/1/012010

Design of Pt-CeO$_x$ hetero-interface on electrodes in polymer electrolyte membrane fuel cells

T Mori,[1]* K Fugane,[1] S Chauhan,[1] M Ito,[1] T Masuda,[1] H Noguchi,[1] and K Uosaki[1]

[1]Global Research Center for Environment and Energy based on Nanomaterials Science (GREEN), National Institute for Materials Science (NIMS), 1-1 Namiki, Tsukuba, Ibaraki, 305-0044 Japan

E-mail:MORI.Toshiyuki@nims.go.jp

Abstract. Pt-CeO$_x$ interface on Pt electrodes was prepared for improvement of both oxygen reduction reaction (ORR) activity of Pt cathode and CO tolerance of Pt anode. The surface of Pt-CeO$_x$ nano-particle/C electrode mainly consists of metallic Pt, ionized Pt (i.e. Pt^{2+}) and Ce^{3+} species. The ORR activity on Pt was improved by the formation of Pt-CeO$_x$ interface. In-situ XAFS analysis suggests that cerium oxide surface in the Pt-CeO$_x$/C was oxidized instead of Pt surface by electrochemical redox reaction of CeO$_x$ at room temperature. This suggests that the Pt-CeO$_x$ interface plays key role for improvement of cathode performance. Also, CO tolerance of Pt was improved by the formation of aforementioned interface. The improvement of CO tolerance of Pt was also observed by using in-situ IR analysis. Based on all experimental data, it is concluded that the design of defect structure in Pt-CeO$_x$ provides us opportunity to make the breakthrough electrodes for fuel cell application.

1. Introduction

Polymer electrolyte membrane fuel cells (PEMFCs) have been developed as clean and efficient power sources for generating electricity by using hydrogen and oxygen. In the case of PEMFCs using Nafion® membrane, the IR loss of electrolyte membrane and anodic over-potential loss are in small level when operation temperature of PEMFCs are under 90°C. In contrast, the over-potential loss at cathode is large because of slow kinetics of oxygen reduction reaction (ORR) on Pt electrode as shown in figure 1(a). If small amount of carbon monoxide (CO) existed in the hydrogen, the over-potential loss on the anode becomes large (figure 1(b)). It is CO poisoning on Pt anode. For development of

* To whom any correspondence should be addressed.

high quality PEMFC devices, aforementioned both slow kinetics of ORR and CO tolerance of Pt have to be improved.

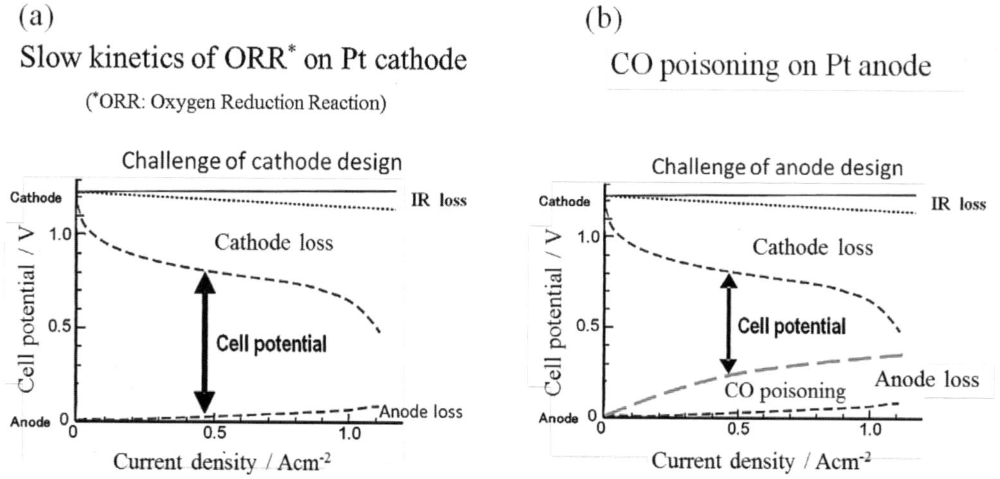

Figure 1. Relationship between cell potential and current density at 80°C; (a): fuel does not include carbon monoxide, (b): fuel includes carbon monoxide more than 3ppm.

For this challenge, the authors focused on the role of heterogeneous interface of Pt and ceria (CeO_x) on Pt electrodes in PEMFCs. The ORR activity on Pt in Pt-CeO_x/conductive carbon (C) was firstly examined by Yu et al.[1] for development of direct methanol fuel cells (DMFCs). 10wt% CeO_x doped Pt/C exhibited the good performance by using single cell of DMFCs. Takahashi et al. [2] reported that the ORR activity on the Pt -CeO_x/C was higher than commercially available Pt/C cathode in the operation condition of PEMFCs. Lim et al. [3] suggested that the active oxygen supplied from CeO_x on Pt surface contributed to the improvement of the ORR activity of the Pt-CeO_x/C cathode.

On the other hand, CO tolerance effect of Pt promoted by CeO_x has been actively investigated because of unique properties of CeO_x such as oxygen storage property, surface basicity, and electrochemical redox reaction between Ce^{4+} and Ce^{3+} in CeO_x at elevated temperature. The overview of proposed mechanism in previously reported papers was given by our short review paper. [4] But the roles of CeO_x in the Pt-CeO_x interface on the cathodes and anodes are still unclear.

In the present work, we try to improve the ORR activity on Pt-CeO_x/C cathode and CO tolerance of Pt in Pt-CeO_x/C anode, also the detail defect structure of Pt-CeO_x interface was characterized by using TEM and XPS. To conclude the defect structural feature well, atomistic simulation was performed by using GULP code, interatomic potential and shell model. Based on those results, we made Pt-CeO_x nanowire/C electrodes and examined the electrochemistry on the electrodes for development of high quality cathode and anode materials.

2. Experimental

Pt-CeO_x nano-sized particle/C electrodes were prepared in a stepwise process. The pure CeO_2 powder was synthesized by a hot ammonium carbonate precipitation method. The starting materials used were commercially available cerium nitrate ($Ce(NO_3)_3$•$6H_2O$, Kanto Chemical Company, Japan) and ammonium bicarbonate ($(NH_4)_2CO_3$, Wako Pure Chemical Industries, Japan) powders. The $Ce(NO_3)_3$•$6H_2O$ powder was dissolved into distilled water. To prepare the fine precipitation, a $(NH_4)_2CO_3$ aqueous solution was heated at 58.0°C in a thermostatic chamber. The $Ce(NO_3)_3$ aqueous solution was dropped into the $(NH_4)_2CO_3$ aqueous solution for 1h period. And the mixture was continuously stirred using magnetic stirrer at 58°C for 24h. After filtration and rinsing, the precipitate was dried at room temperature in a N_2 gas flow. The dried powders were calcined at 400 °C for 2h in an O_2 gas flow. The CeO_x nanowire was synthesized by using an alcohol thermal method. The starting materials used were commercially available cerium chloride ($CeCl_3 \cdot H_2O$, Nacalai Tesque Company, Japan), cetyltetraethylammonium bromide ($C_{19}H_{42}BrN$, Wako Pure Chemical Industries, Ltd., Japan) and urea ($(H_2N)_2CO$, Nacalai Tesque Company, Japan) powders. Powders were dissolved in dehydrated ethanol and mixed using a magnetic stirrer at room temperature for one hour under a dry nitrogen atmosphere. $C_{19}H_{42}BrN$ and $(H_2N)_2CO$ were respectively used as surfactant for micelle formation and as base for control of pH during the reaction. The reaction mixture was filtered by ultrafiltration and half the filtrate was transferred to a 25ml Teflon autoclave. CeO_x nanowire synthesis was performed by heating the 25ml Teflon autoclave at 80°C for 4 days in a heating oven. After filtration and rinsing, the mixture was dried at room temperature in a N_2 gas flow. To impregnate fine Pt particles on pure CeO_2 particles, hexachloroplatinic(IV) acid (H_2PtCl_6•$6H_2O$) powder was dissolved in ethanol at the glove box where the moisture level was kept below 10ppm. The previously prepared pure CeO_2 powders were dispersed in the H_2PtCl_6 ethanol solution and dried in the glove box. The dried mixture was calcined at 400 °C for 1h in a 10% H_2/He gas flow. For preparation of Pt-CeO_x nanowire/C electrodes, potassium hexachloroplatinate (IV) ($K_2PtCl_4 \cdot H_2O$, Sigma-aldrich Ltd., Japan) powder was dissolved in distilled water and stirred for one hour. The aqueous K_2PtCl_4 solution was added to a suspension of the previously prepared CeO_x nanowires dispersed in distilled water. The resulting mixture was left standing at room temperature for 12h. The platinum complex in the mixture was reduced at room temperature for 1h in aqueous sodium borohydride ($NaBH_4$, Sigma-aldrich Ltd., Japan) solution. The mixture was rinsed with water and ethanol, and then it was dried at room temperature in nitrogen atmosphere. In the final stage, the prepared Pt-CeO_x and C were dispersed in ethanol and dried in a N_2 gas flow.

The crystal phases of Pt-CeO_x/C were identified by selected area electron diffraction analysis and X-ray diffraction analysis. The particle size and particle morphology were measured by TEM. The TEM observation was performed by using JEM-2000EX/T (JEOL, Co. Ltd, Japan) electron microscope operated at 200 kV. The surface chemical states of Pt-CeO_x/C were examined by using hard X-ray photoelectron spectroscopy (HXPS) for average chemical states analysis of the interface and soft X-ray photoelectron spectroscopy (SXPS) for chemical states analysis of the top surface.

To characterize the CO poisoning phenomena on Pt anode, in-situ FT-IR analysis was performed. The electrolyte solution used was the 0.5M H_2SO_4 solution saturated with Ar for 10min. Pure CO gas was purged into the electrolyte solution at 0.07V for 30min to obtain an equilibrium adsorption of CO. CO in the bulk solution was then removed by purging with Ar while keeping 0.07V for 5min. IR measurement was carried out from 0.07 to 1.07V. The potential shifts were every 50mV interval. And the spectral resolution used was 2cm^{-1}.

The cathode property of Pt-CeO_x/C was examined by linear voltammetry in 0.5M H_2SO_4 aqueous solution in the potential ranging from 0.95 to 0 V (V vs. Ag/AgCl). Anode properties of Pt-loaded CeO_x nanowire mixed with C were examined by cyclic voltammetry (CV) in mixed aqueous solution of 0.5M CH_3OH and 0.5M H_2SO_4 in the potential ranging from 0 to 1.5V (vs. Ag/AgCl). The electrochemical measurements were carried out using a standard three-electrode glass cell with a rotating disk electrode after that had a pre-treatment by heating at a 28 °C. A Pt foil and Ag/AgCl were used as a counter and reference electrodes, respectively. The measured potentials were converted to the reversible hydrogen electrode (RHE) scale. Prior to the electrochemical measurements, the electrochemical pre-treatment for conditioning of the anode surface was carried out by means of 30 cycle sweeps in the potential range from 0 to 1.5V (vs. Ag/AgCl).

3. Results and discussion

The TEM photograph of as prepared Pt-CeO_x/C is presented in figure 2. On the selected area electron diffraction pattern (SAEDP) taken from as prepared samples, the diffraction rings belong to metallic platinum and fluorite structured ceria. The Pt particles were mainly situated on well crystalline ceria particles and Pt particles partially embedded into ceria particle. [5]

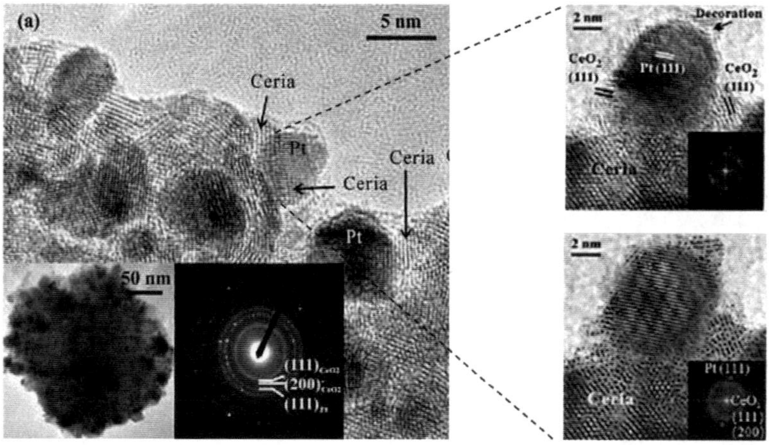

Figure 2. TEM photograph and selected area electron diffraction pattern (SAEDP) taken from as prepared sample. (a): general image and SAEDP. (b) Magnified image and FFT patterns.

However, the chemical composition and general view of Pt-CeO_x/C electrode were conspicuously changed after the electrochemical pre-treatment for conditioning of the electrode surface. As shown in figure 3, small amount low CeO_x partially covered the Pt surface. On SAEDP taken from this sample, the careful analysis of SAED patterns revealed that the short-range ordered materials and nano-crystallites in the residual oxides could include fluorite-structured ceria (CeO_2, d(111) ≈ 3.1 Å),

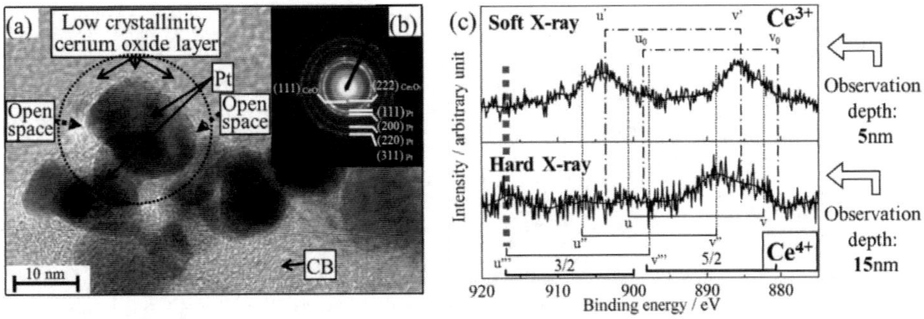

Figure 3. TEM photograph (a), selected area electron diffraction pattern (b) and Ce3d spectra taken from electrochemically pre-treated Pt-CeO_x/C electrode.

reduced ceria (Ce_2O_3, d(111) ≈3.2-3.4 Å), PtO, and other compounds that were difficult to identify at this stage. Those microanalysis results suggest that free CeO_2 without Pt-CeO_x interaction is resolved into H_2SO_4 aqueous solution after 30 cycle sweeps in the potential range from 0 to 1.5V (vs. Ag/AgCl). It is concluded that Schottky type Pt_{Ce}''- $2Vo^{\bullet\bullet}$-Ce_{Ce}' defect clusters were formed in the Pt-CeO_x interface in our preparation process. [6] We expect that the electrode performance of Pt-CeO_x/C will be enhanced by the formation of aforementioned heterogeneous Pt-CeO_x interface on Pt.

The ORR activity observed for conventional Pt/C and Pt-CeO_x/C were examined by using Tafel plots. The intersect point of Tafel line and potential axis in the Tafel plots reflects the level of exchange current density of ORR on the electrodes. If aforementioned intersect point shifted in the high potential region, the ORR activity is in high level. After the electrochemical pre-treatment, the ORR activity observed for Pt/C was slightly improved by enhancement of surface area of Pt.

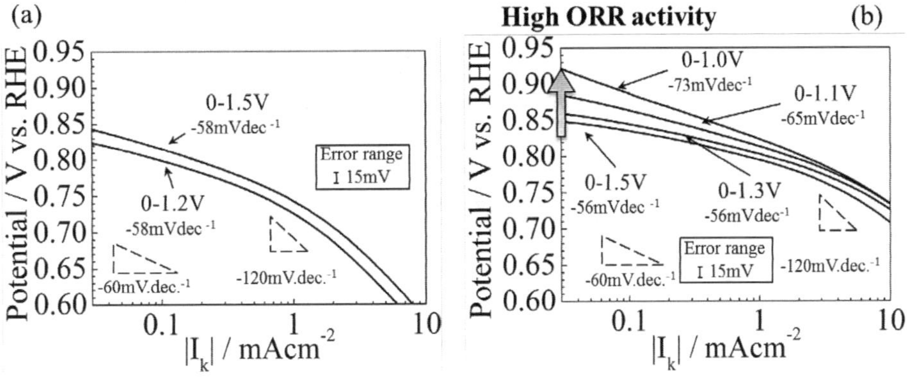

Figure 4. Tafel plots obtained from conventional Pt/C (a) and Pt-CeO_x/C (b) electrodes.

In contrast, the ORR activity on Pt-CeO_x/C was conspicuously improved after the electrochemical pre-treatment. [7,8] Also, aforementioned intersect point observed for electrochemically pre-treated Pt-CeO_x/C was much higher than that of conventional Pt/C. This clearly indicates that the ORR activity on Pt in Pt-CeO_x/C was improved by the formation of Pt-CeO_x hetero-interface.

To conclude why the surface of Pt in Pt-CeO_x/C is active in cathode reaction, in-situ electrochemical X-ray absorption fine structure (XAFS) analysis was performed by using home-made electrochemical cell in synchrotron beam line (BL14B1 of SPring-8). In-situ XAFS measurements at the Pt L_3 and Ce L_3 absorption edges of the Pt−CeO_x/C catalyst suggested that oxide formation on Pt was suppressed by the presence of CeO_x as Ce^{3+} was oxidized to Ce^{4+} instead of Pt surface in cathodic condition. The electrochemistry observed for Pt-CeO_x/C agreed with this result well. Based on those results, we conclude that the inhibition of Pt oxide formation is considered to be the primary factor for enhancement of the ORR activity on Pt in Pt-CeO_x/C because ORR activity of the Pt oxide surface is much lower than that at the bare Pt surface. [9]

Since the formation of defect structure in Pt-CeO_x interface is key for design of high quality cathode, the CeO_x nanowires instead of CeO_x nanoparticles were prepared by using alcohol thermal method. [10] SEM micrographs shown in figure 5 indicate that the product was composed of only nanowires. Also the average diameter of CeO_x nanowires was approximately 35 nm. The aspect ratio of the fabricated CeO_x nanowires was greater than 100. Electron diffraction rings can be assigned to CeO_2 (i.e. CeO_2 (d(111) ≈ 3.1 Å, d(200) ≈2.7 Å, d(220) ≈1.9 Å)) in the SAEDP on SEM image of figure 5. We concluded that the nano-scale surface roughness of CeO_x nanowires makes them suitable for preparation of functional interfaces between nanosized Pt and CeO_x nanowire surfaces. After the preparation of CeO_x nanowaires, Pt-CeO_x nanowire/C electrode was prepared. The surface analysis of

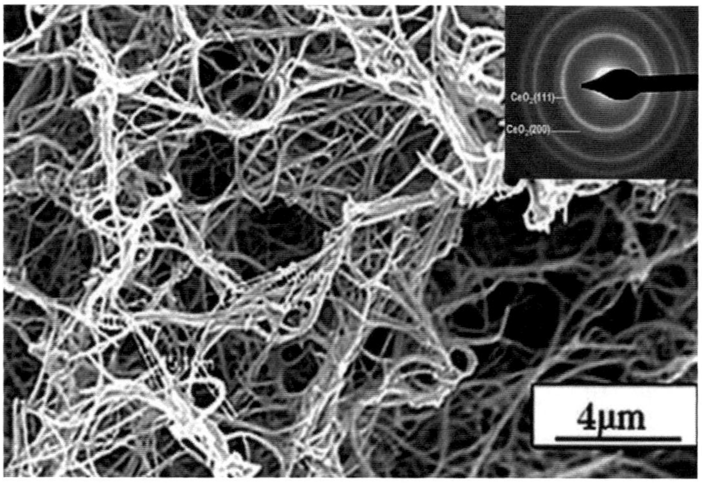

Figure 5. SEM photograph and SAEDP taken from CeO_x nanowire.

electrochemically pre-treated Pt-CeO_x nanowire/C suggests that the electrode surface mainly consists of Pt (i.e. metallic Pt), ionized Pt (i.e. Pt^{2+}) and Ce^{4+} species. This suggests that the Pt-CeO_x interface observed for Pt-CeO_x nanowire/C is different from the case of Pt-CeO_x nanopartcile/C. The Frenkel type defect structure would be formed in the Pt-CeO_x interface on Pt-CeO_x nanowire/C electrode. After 300 cycle sweeps in electrochemical pre-treatment process, the stationary state condition was observed in the case of Pt-CeO_x nanowire/C. The ORR activity was observed by using 5wt% Pt-CeO_x

IOP Conf. Series: Materials Science and Engineering **54** (2014) 012010 doi:10.1088/1757-899X/54/1/012010

nanowire/C. The intersect point of Tafel slop and potential axis at 0.3 $mAcm^{-2}$ in Tafel plot observed for 5wt% Pt-CeO_x nanowire/C was almost same to aforementioned intersect point observed for 20wt% Pt-CeO_x nanoparticle/C cathode. This suggests that we can reduce the Pt content from 20wt% to 5wt% in our Pt-CeO_x cathode by enhancement of ORR activity on our electrode.

For development of our design concept regarding metal – oxide hetero-interface on electrodes in fuel cell reaction, we examined the electro-oxidation reaction of methanol on conventional Pt/C and Pt-CeO_x nanowire /C anodes as well as ORR activity on cathodes. [10] The estimated onset potential of the commercially available 20wt% Pt/C anode was approximately 0.5 V (vs. RHE). This is similar to the previously reported value observed for Pt/C (i.e. 0.5 V vs. RHE). [11] In contrast, the methanol electro-oxidation on 5wt% Pt-loaded CeO_x nanowire/C commenced from approximately 0.19 V (vs. RHE), which is much lower than that of the aforementioned commercially available Pt/C. Also, this onset potential observed for Pt–CeO_x nanowire/C was conspicuously lower than previously reported PtRu–CeO_x nano-particle/C (0.30 V vs. RHE) [12] under similar experimental condition. Those data clearly indicates that CO tolerance of Pt in Pt–CeO_x/C can be improved by the formation of Pt-CeO_x interface on Pt anode.

To examine the influence of the CO poisoning phenomenon on the formation of Pt–CeO_x interface, CO desorption behaviour from Pt surface under applied potential was observed by using in-situ IR analysis method (figure 6). The absorption band at 2106 cm^{-1} corresponds to the adsorbed CO on Au electrode. Alternatively, the strong absorption bands at 2030cm^{-1} and 2012 cm^{-1} in figure 6(a) and 2046cm^{-1} and 2017 cm^{-1} in figure 6(b) were assigned to the linear bonded CO on Pt. The bridged (doubly coordinated) CO on Pt was observed at 1842cm^{-1} in 1856cm^{-1} in this figure. In the present analysis, at 70mV (vs.RHE), the center of band spectra for adsorbed CO on Pt in Pt-CeO_x/C conspicuously shifted in higher wavenumber region as compared with those for adsorbed CO on Pt in the Pt/C. It suggests that the adsorbed strength of CO molecule on Pt surface becomes weak by formation of Pt-CeO_x interface on Pt. This supported the results of observation of onset potential of methanol oxidation. In addition, our atomistic simulation suggests that the defect structures in the interfaces on Pt-CeO_x nano-particles/C and Pt-CeO_x nanowire/C mainly consists of Schottky type defect (i.e. Pt_{Ce}''- $2Vo^{\bullet\bullet}$-Ce_{Ce}') and Frenkel type defect (i.e. $2Pt_i^{\bullet\bullet}$-$4O_i''$-$4V_o^{\bullet\bullet}$-V_{ce}''''), respectively. In the present manuscript, Kröger–Vink notation is used for simple

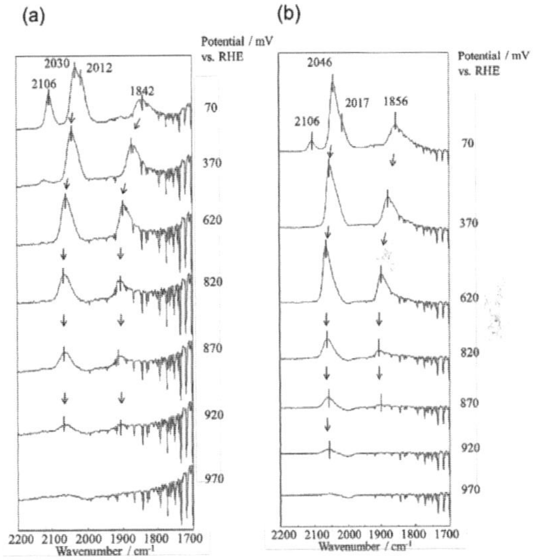

Figure 6. In-situ FT-IR spectra taken from conventional Pt/C (a) and Pt-CeO_x/C (b).

explanation of defect formation. Moreover, our atomistic simulation suggests that aforementioned clusters were just building unit for development of large defect clusters in the Pt-CeO_x interface on Pt electrode. Also, our TEM analysis supported the results of atomistic simulation in the present work.

Based on all experimental results, we concluded that defect interface structure design in the interface is quite important for development of anode and cathode for fuel cell application.

4. Conclusion

For improvement of ORR activity on Pt and CO tolerance of Pt, Pt-CeO_x/C cathodes and anodes were prepared by a combined process of precipitation and impregnation method. Also, the Pt-CeO_x interface was characterized by using TEM, XPS, in-situ XAFS and in-situ IR. To conclude the detail defect structure of interface, atomistic simulation was performed. Based on all experimental results, it is concluded that the present work is promising to provide both a rational strategy for design of metal-oxide interface which promotes electrode performance and high quality cathode and anode for fuel cells.

References

[1] Yu H B, Kim J H, Lee H I, Scibioh M A, Lee J, Han J, Yoon S P and Ha H Y, 2005 *J. Power Sources*, **140** 59-65.

[2] Takahashi M, Mori T, Yoshikawa H, Togasaki H, Fugane K, Tada A, Matolin V and Drennan J, 2008 *Transactions of the Materials Research Society of Japan*, **33** 1101-04.

[3] Lim D H, Lee W D, Choi D H, Lee H I, 2010 *Appl. Catal. B: Environ*, **94** 85-96.

[4] Mori T, Ou D R, Zou J, Drenna J, 2012 *Progress in Natural Science: Materials International*, **22** 561-571.

[5] Ou D R, Mori T, Togasaki H, Takahashi M, Ye F, and Drennan J, 2011 *Langmuir*, **27**, 3859-3866.

[6]Ou D R, Mori T, Fugane K, Togasaki H, Ye F, and Drennan J, 2011 *Journal of Physical Chemistry C*, **115**, 19239-19245.

[7] Fugane K, Mori T, Ou D R, Suzuki A, Yoshikawa H, Masuda T, Uosaki K, Yamashita Y, Ueda S, Kobayashi K, Okazaki N, Matolinova I and Matolin V, 2011 *Electrochimica Acta*, **56**, 3874-3883.

[8] Fugane K, Mori T, Ou D R, Yan P F, Ye F, Yoshikawa H, and Drennan J,2012 *Langmuir*, **28**, 16692-16700.

[9] Masuda T, Fukumitsu H, Fugane K, Togasaki H, Matsumura D, Tamura K, Nishihata Y, Yoshikawa H, Kobayashi K, Mori T, and Uosaki K, 2012 *Journal of Physical Chemistry C*, **116**, 10098–10102.

[10] Chauhan S, Richards G J, Mori T, Yan P F, Hill J P, Ariga K, Zou J and Drennan J, 2013 *Journal of Materials Chemistry A*, **1**, 6262-6270.

[11] Vel´azquez-Palenzuela A, Francesc C, Garrido J A, Arias C, Rodriguez R M, Brillas E, and Cabot P L, 2011 *J. Power Sources*, **196**, 3503-3512.

[12] Huang S Y, Chang C M, and Yeh C T, 2006 *J.Catal.*, **241**, 400-406.

Reactive sintering of pseudobrookite-type $MgFeNbO_5$ and $MgFeTaO_5$ ceramics

Yoshikazu Suzuki[1] and Hiroya Abe[2]

[1]Faculty of Pure and Applied Sciences, University of Tsukuba,
1-1-1 Tennodai, Tsukuba, Ibaraki, 305-8573, Japan
[2]Joining and Welding Research Institute, Osaka University,
11-1 Mihogaoka, Ibaragi, Osaka 567-0047, Japan

E-mail: suzuki@ims.tsukuba.ac.jp

Abstract. Pseudobrookite-type ceramics have relatively low bulk thermal expansion among oxides and are expected to be a new family of functional materials. $MgFeNbO_5$ and $MgFeTaO_5$, Fe-containing pseudobrookite, were synthesized by a reactive sintering method. Solid-state reaction behavior was studied in detail by high-temperature X-ray diffraction (HT-XRD).

1. Introduction

Pseudobrookite-type ceramics (orthorhombic structure, generally expressed as Me_3O_5) have relatively low bulk thermal expansion among oxides, which is caused by the strong crystallographic anisotropy (Figure 1) and the resulting microcracks [1-4]. Among the pseudobrookite-type ceramics, Al_2TiO_5 is well known and commercialized as a structural material for cyclic heating/cooling environment. As for a non-Al_2TiO_5 phase, our group has recently reported the synthesis of $MgTi_2O_5$-based uniformly porous ceramics with 3-D network structure (UPC-3D). $MgTi_2O_5$ has less crystallographic anisotropy than Al_2TiO_5, and $MgTi_2O_5$ is expected to exhibit well-balanced properties [5-7].

In order to further expand the feasibility of the pseudobrookite-type ceramics toward functional materials, substitution strategy will be effective from analogy with perovskite-type ceramics and so on. In an early work by Bayer in 1971 [1], Fe-containing compounds, $MgFeNbO_5$ and $MgFeTaO_5$, were synthesized by solid-state reactions at 1250°C. They can be derived from $MgTi_2O_5$ by substituting Fe^{3+} + Nb^{5+} (or Ta^{5+}) for 2 Ti^{4+}. Bayer also reported that attempts to prepare other $M^{2+}\,M^{3+}\,M^{5+}\,O_5$ compounds were unsuccessful, *e.g.*, the following combinations gave no pseudobrookite but mixture of rutile, columbite and spinel phases: $MgAlNbO_5$, $MgAlTaO_5$, $MgGaNbO_5$, $MgGaTaO_5$, $MgCrNbO_5$, $MgCrTaO_5$, $MgFeSbO_5$, $MgFeVO_5$ and the corresponding compositions with Zn, Ni, Co, and Cu instead of Mg [1]. Successful synthesis of $MgFeNbO_5$ and $MgFeTaO_5$ can be substantially attributable to the similar ionic radii of Ti^{4+}, Fe^{3+}, Nb^{5+} and Ta^{5+}, as summarized in Table 1 [8]. In spite of their potentiality as new functional materials, there were few reports on $MgFeNbO_5$ and $MgFeTaO_5$ (except ICDD-JCPDS database based on the Bayer's early work).

In this paper, we have synthesized $MgFeNbO_5$ and $MgFeTaO_5$ by a reactive sintering method using solid-state reactions. High-temperature X-ray diffractometry (HT-XRD) was used to analyze the *in situ* reaction behavior. Microstructure of bulk $MgFeNbO_5$ and $MgFeTaO_5$ will be also reported.

[1] To whom any correspondence should be addressed.

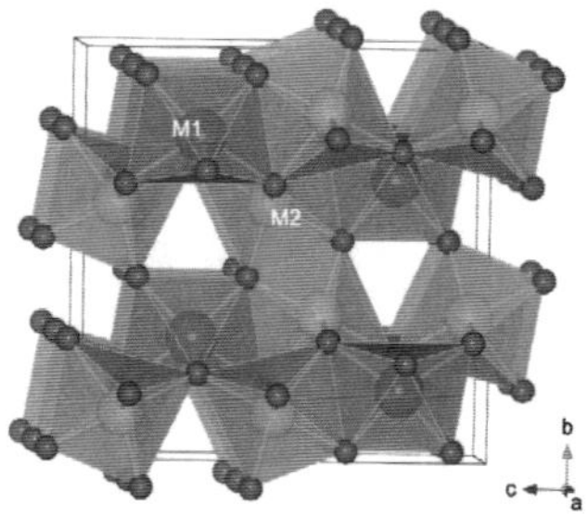

Figure 1. Crystal structure of $MgTi_2O_5$ (in the *Cmcm* (63) space group notation).

Table 1. Reported ionic radii [8].

	Goldschmidt	Pauling	Ahrens
Mg^{2+}	0.78	0.65	0.66
Ti^{4+}	0.64	0.68	0.68
Fe^{3+}	0.67		0.64
Nb^{5+}		0.70	0.69
Ta^{5+}			0.68
O^{2-}	1.32	1.40	

2. Experimental

2.1. Sample preparation
Commercially available fine powders, $MgCO_3$ (basic) (99.9%, Kojundo Chemical Laboratory Co. Ltd., Saitama, Japan), α-Fe_2O_3, Nb_2O_5 and Ta_2O_5 (99.9% each, Wako Pure Chemical Ind., Ltd, Osaka, Japan), were used as the starting materials. The commercial $MgCO_3$ (basic) powder used in this study was actually composed of hydromagnesite phase, $Mg_5(CO_3)_4(OH)_2$•$4H_2O$ ($4MgCO_3$• $Mg(OH)_2$•$4H_2O$), and thus, TG-DTA analysis was used to determine the actual MgO fraction prior to the mixing [7]. The $MgCO_3$ (basic), α-Fe_2O_3 and Nb_2O_5 (or Ta_2O_5) powders of Mg : Fe : Nb (or Ta) = 1 : 1 : 1 in mole fraction were mixed using agate mortar for 1 h using ethanol. Due to the heaviness (high specific gravity) of Nb_2O_5 and Ta_2O_5 powders, the mixing in a mortar instead of the ball milling was used in this study.

2.2. HT-XRD for the powder mixtures
The reaction behavior of the $MgCO_3$(basic)-Fe_2O_3-Nb_2O_5 (or Ta_2O_5) mixed powder (without compressing) was analyzed by HT-XRD (Multiflex, Cu-K_α, 40 kV and 40 mA, XRD, RIGAKU, Japan) with a vertical-type goniometer, in the range between room temperature and 1300°C in air. The ramp rate of the furnace was 10°C/min. Each XRD pattern was acquired after 5 min holding at each temperature. The scanning rate was 2θ=4°/min and the scanning range was 2θ=10-60°, hence, 5 min holding + 15 min scanning at each temperature. Since the Bragg-Brentano geometry was used in this study, change of the sample height by shrinkage (or sintering) made a little error of peak positions at high temperatures (typically $\Delta 2\theta \sim 0.1$-0.2°).

2.3. Preparation and evaluation of bulk ceramics
The mixed powders (without any binder) were mold-pressed (~16 MPa) with dimensions of 15 mm in diameter and ~3 mm in thickness, and then cold isostatically pressed at 200 MPa to obtain the green compacts. They were sintered in air at 1250°C for 2 h. The constituent phases of the bulk ceramics were analyzed by XRD (Multiflex, Cu-K_α, 40 kV and 40 mA). Prior to the powder XRD measurement, bulk pellets were pulverized. The microstructure of the bulk ceramics was characterized by scanning electron microscopy (SEM, SU-70, Hitachi High-Tech., Tokyo, Japan).

3. Results and disucussion

3.1. High-temperature XRD analysis for solid state reactions
Figure 2 shows HT-XRD patterns for (a) $MgCO_3$ (basic)-Fe_2O_3-Nb_2O_5 powder and (b) $MgCO_3$ (basic)-Fe_2O_3-Ta_2O_5 powder. Top patterns in both figures were taken at room temperature after cooling.

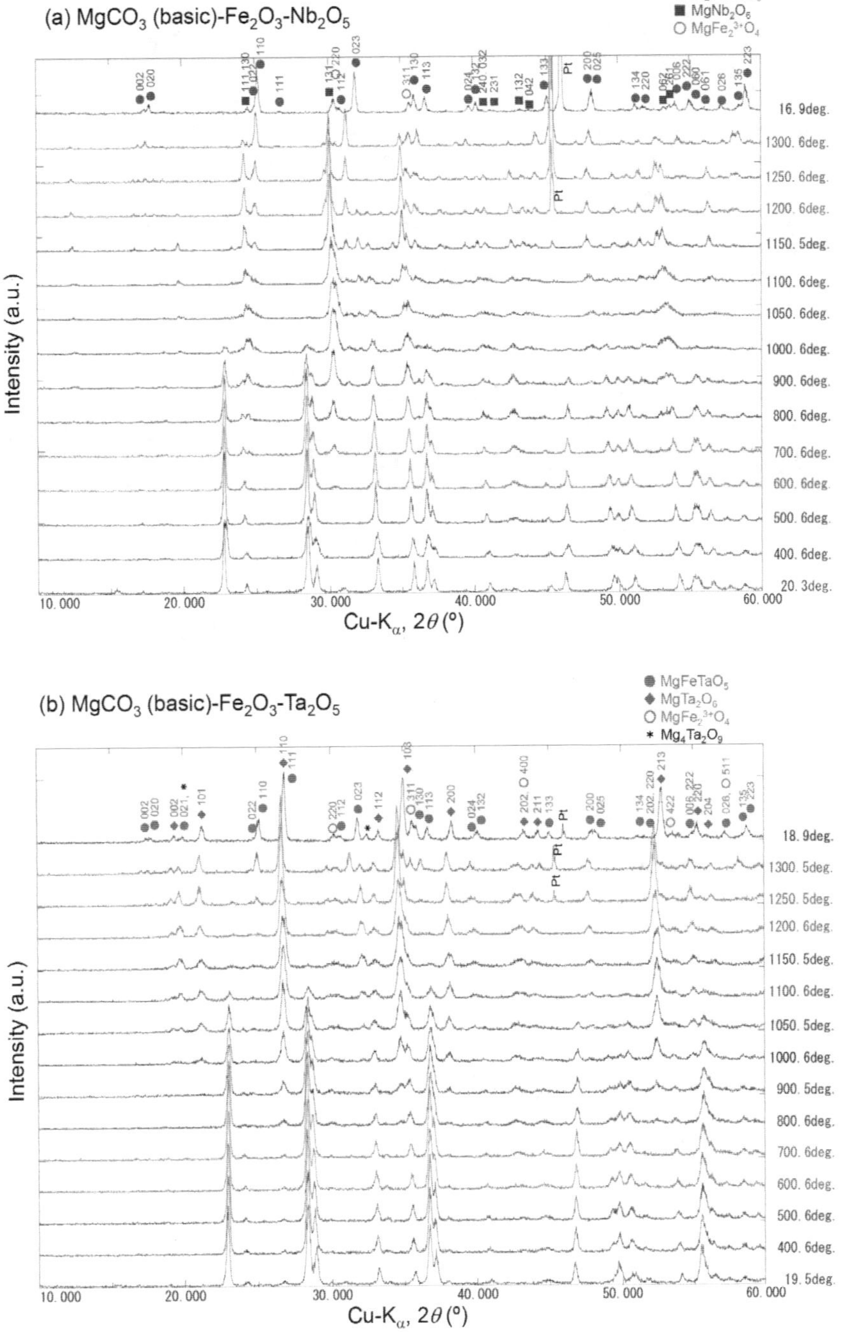

Figure 2. High-temperature *in situ* XRD patterns for (a) $MgCO_3$ (basic)-Fe_2O_3-Nb_2O_5 powder and (b) $MgCO_3$ (basic)-Fe_2O_3-Ta_2O_5 powder. Mg : Fe : Nb (or Ta) = 1 : 1 : 1 in mole fraction. Top patterns in both figures were taken at room temperature after cooling.

As for the Nb-system (Fig. 2(a)), formation of binary compounds, such as $MgNb_2O_6$ (magno-columbite) and spinel-type $MgFe_2^{3+}O_4$ (magnesioferrite), were confirmed as intermediate phases. $MgFeNbO_5$ started to form around 1100°C. At 1300°C and after cooling, the main component was $MgFeNbO_5$. Even with neither the compression of the powder nor long-term heating, $MgFeNbO_5$ formed easily in the HT-XRD measurement. At ≥1150°C, Pt peak at ~45-46° from the sample stage became strong, which demonstrated the shrinkage (sintering) at high temperatures.

As for the Ta-system (Fig. 2(b)), formation of binary compounds, such as $MgTa_2O_6$ and spinel-type $MgFe_2^{3+}O_4$, were also identified as intermediate phases. Note that the Ta-containing phases generally give high X-ray intensity due to the large atomic scattering factor of Ta. At ≥1250°C, the formation of $MgFeTaO_5$ was confirmed and small Pt peak at ~45-46° appeared. The smaller Pt peak from sample stage indicated that $MgFeTaO_5$ phase did not so sinterable than $MgFeNbO_5$.

3.2. *Appearance and XRD analysis of bulk ceramics*

Figure 3 shows the appearance and XRD patterns of bulk ceramics sintered at 1250°C for 2 h. The bulk density (determined by dimension and mass) of $MgFeNbO_5$ and $MgFeTaO_5$ pellets were 3.33 g/cm^3 and 2.84 g/cm^3, respectively. Since the reported (X-ray) density of $MgFeNbO_5$ and $MgFeTaO_5$ are 4.375 g/cm^3 [9] and 5.876 g/cm^3 [10], the relative densities of these pellets are estimated to be ~76 % for $MgFeNbO_5$ and ~48% for $MgFeTaO_5$. In good agreement with the HT-XRD analysis, $MgFeTaO_5$ was less sinterable than $MgFeNbO_5$.

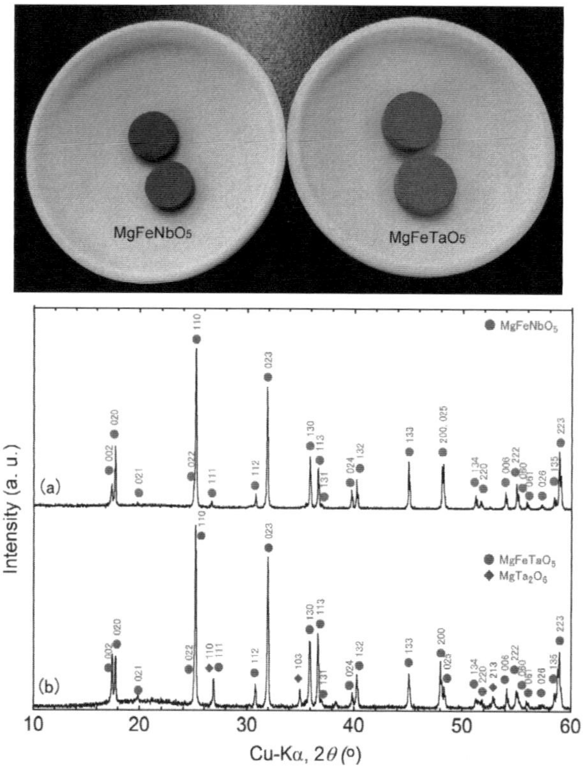

Figure 3. Appearance and XRD patterns of bulk ceramics sintered at 1250°C for 2 h. Dark brown for $MgFeNbO_5$ pellets and ocher for $MgFeTaO_5$.

All peaks are well-indexed in the XRD analysis. For the Nb-system, substantially pure $MgFeNbO_5$ was successfully obtained. On the other hand, for the Ta-system, $MgFeTaO_5$ as well as a little $MgTa_2O_6$ were found in the bulk pellets. Higher reactivity in the Nb-system can be explained by the formation of some eutectic liquid at high temperatures; lower reactivity in the Ta-system can be attributed to the refractory nature of Ta-containing compounds with higher melting points, and thus, less liquid-phase formation resulted in porous microstructure. To obtain dense and pure $MgFeNbO_5$ and $MgFeTaO_5$, crushing and re-sintering must be effective.

3.3. Microstructure

Microstructure of the pellet surface is given in Figure 4. In good agreement with the above discussion, the Nb-system had higher density and larger grain size than the Ta-system. On the surface of the $MgFeNbO_5$ (figure 4(a)), plenty of curved steps were observed, which suggests the liquid-phase formation. Similarly to other pseudobrookite ceramics, some microcracks existed in the bulk ceramics.

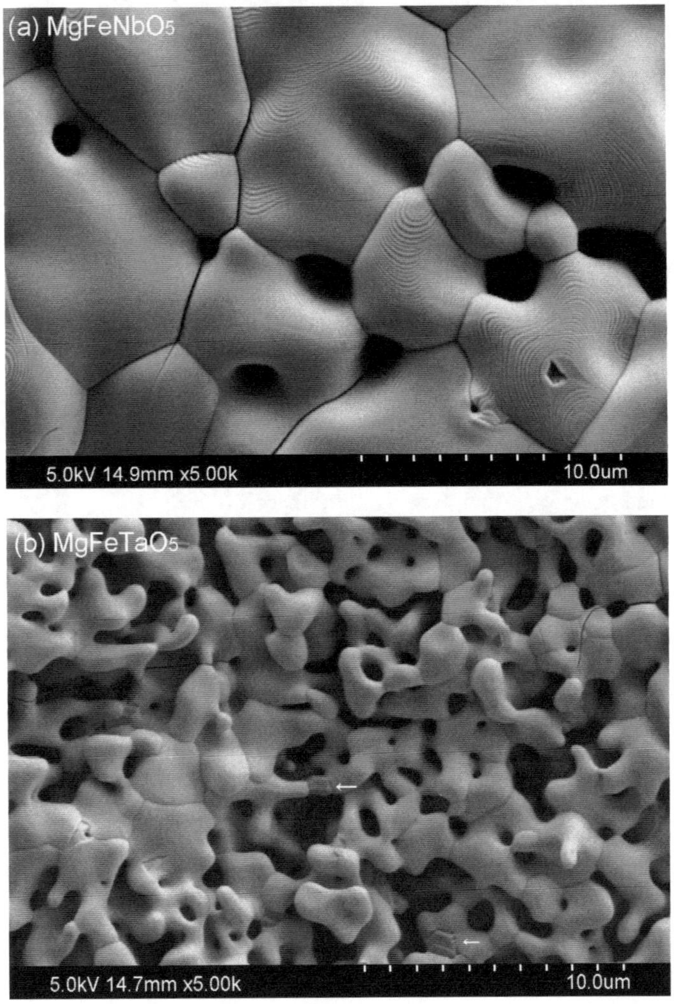

Figure 4. SEM images of the surface structure of bulk ceramics sintered at 1250°C for 2 h. (a) $MgFeNbO_5$ and (b) $MgFeTaO_5$.

The Ta-system has open porous microstructure with 3-D network, caused by the gas emission from $MgCO_3$ (basic), less liquid phase formation, and its less sinterability than the Nb-system. The faceted grains with arrows (figure 4(b)) presumably correspond to the remnant $MgTa_2O_6$.

4. Conclusions

In this paper, we have synthesized $MgFeNbO_5$ and $MgFeTaO_5$ by a reactive sintering method using solid-state reactions. HT-XRD revealed that the reaction behavior of $MgCO_3$ (basic)-Fe_2O_3-Nb_2O_5 and $MgCO_3$ (basic)-Fe_2O_3-Ta_2O_5 mixed powders. $MgNb_2O_6$ (or $MgTa_2O_6$) and spinel-type $MgFe_2^{3+}O_4$ were formed as binary intermediate phases. Ternary $MgFeNbO_5$ formed at ~1100°C, and $MgFeTaO_5$ formed at ~1250°C. Reactive sintering of these powders at 1250°C for 2 h gave pure $MgFeNbO_5$ (relative density of ~76 %) and $MgFeTaO_5$ with some remnant $MgTa_2O_6$ (relative density of ~48 %). SEM observation suggested the formation of liquid phase during the sintering of $MgFeNbO_5$.

Acknowledgement

This work was supported by Grant-in-Aid for Science Research No. 23350111 for Basic Research: Category B and Joint Research Project of JWRI, Osaka University. We thank Mr. Kenji Toumoto at JWRI for his help on SEM observation. AtomWork database (NIMS) [11] and Vesta Ver. 2 software (Momma and Izumi [12]) were used for crystal drawing.

References

[1] G. Bayer, J. Less-Common Metals, **24**, 129 (1971).
[2] J. J. Cleveland, R. C. Bradt, J. Am. Ceram. Soc., **61**, 478 (1978).
[3] D. Taylor, British Ceram. Trans. J., **86**, 1 (1987).
[4] Y. Suzuki, Y. Shinoda, Sci. Tech. Adv. Mater., **12**, no.034301 (2011).
[5] Y. Suzuki and M. Morimoto, J. Ceram. Soc. Jpn., **118**, 819 (2010).
[6] Y. Suzuki and M. Morimoto, J. Ceram. Soc. Jpn., **118**, 1212 (2010).
[7] Y. Suzuki, T. S. Suzuki, Y. Shinoda, and K. Yoshida, Adv. Eng. Mater., **14**, 1134 (2012).
[8] F. Galasso, "Structure and Properties of Inorganic Solids," Pergamon Press Ltd., Oxford 1970.
[9] ICDD-JCPDS database, no. 24-0707 for $MgFeNbO_5$, 1974.
[10] ICDD-JCPDS database, no. 25-0522 for $MgFeTaO_5$, 1975.
[11] NIMS AtomWork Database, http://crystdb.nims.go.jp/
[12] K. Momma and F. Izumi, J. Appl. Crystallogr., **41**, 653 (2008).

Tsukuba International Conference on Materials Science 2013 IOP Publishing
IOP Conf. Series: Materials Science and Engineering **54** (2014) 012012 doi:10.1088/1757-899X/54/1/012012

Ru-catalyzed site-selective direct arylation polycondensation via *ortho*-metalation of pyrrole derivative

T Kanbara, J Kuwabara and W Lu

Tsukuba Research Center for Interdisciplinary Materials Science (TIMS), Graduate School of Pure and Applied Science, University of Tsukuba, 1-1-1 Tennodai, Tsukuba 305-8573, Japan

E-mail: kanbara@ims.tsukuba.ac.jp

Abstract. A π-conjugated polymer consisting of pyrrole and fluorene units was synthesized via Ru-catalyzed direct arylation polycondensation. The introduction of a 2-pyrimidinyl substituent into the *N*-position of the pyrrole monomer as a directing group induced *ortho*-metalation, with effective site-selective polycondensation at the α-position of the pyrrole monomer, without the need for protection of the β-position. Subsequent removal of the 2-pyrimidinyl substituent led to the elongation of π-conjugation along the main chain of the polymer via elimination of the steric hindrance due to the bulky substituent.

1. Introduction

π-Conjugated polymers have been extensively investigated for utilization in the fabrication of a variety of optoelectronic devices such as organic photovoltaic cells, field effect transistors, and light emitting diodes [1–5]. These polymers have been predominantly synthesized by polycondensation using transition metal-catalyzed cross-coupling reactions such as the Suzuki–Miyaura and Migita–Kosugi–Stille cross-coupling reactions [6–12]. Alternatively, syntheses of π-conjugated polymers via a dehydrohalogenative cross-coupling reaction (scheme 1), so-called direct arylation, have recently been actively investigated as this protocol should provide significant benefits by decreasing the number of reaction steps and reducing undesired waste from the organometallic reagents [13–39].

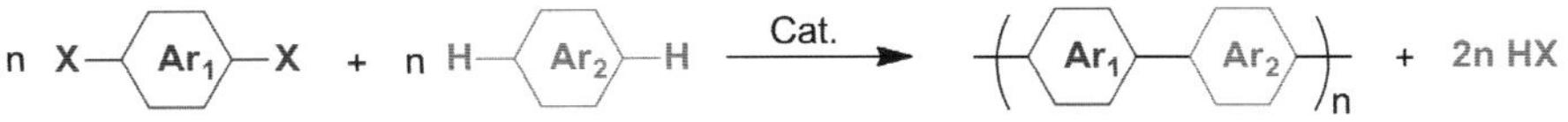

Scheme 1.

Direct arylation polycondensation, however, has some limitations: C–H activation at undesired positions of the aromatic monomers provides branched and cross-linked structures in the polymer chains [20–22,26–28,29–33], which affects the solubility and optical properties of the polymers [21,28]. We previously reported that Pd-catalyzed direct arylation polycondensation of 3,3′,4,4′-tetramethylbithiophene with dibromoarylenes gave the corresponding polymers with high molecular weights and good solubility. In contrast, the use of 2,2′-bithiophene as a monomer gave insoluble products due to the occurrence of direct arylation at the β-position of the thiophene unit [32]. The β-position side reaction was also confirmed by carrying out a model reaction of bithiophene with *p*-

bromotoluene (scheme 2), where a trace amount of trisubstituted bithiophene was observed. Consequently, this minor side reaction induced the formation of cross-linked structures during the polycondensation reaction.

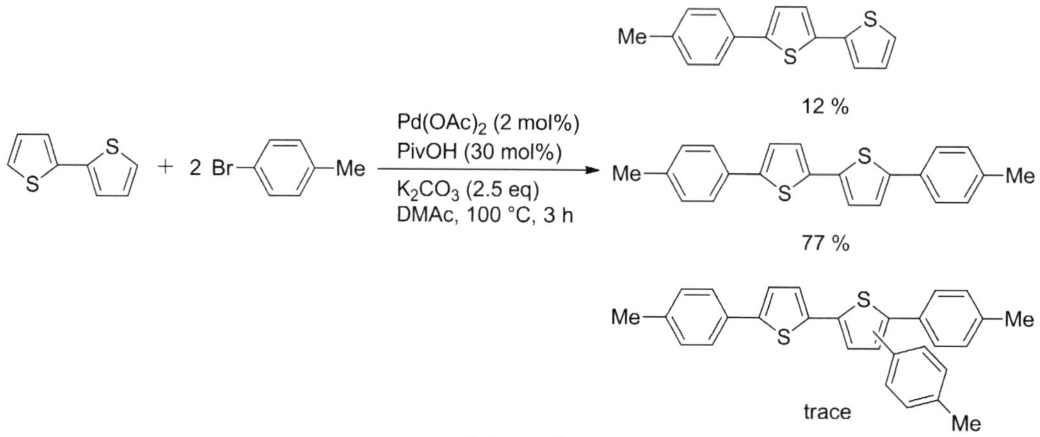

Scheme 2.

These observations initiated our interest in achieving direct arylation polycondensation with site selectivity in order to suppress the undesired side reaction. We thus focused on Ru-catalyzed direct arylation of pyrrole derivatives bearing a directing group [40,41]. In this reaction, the presence of the directing group on the aromatic substrates induced *ortho*-selective C–H bond functionalization via *ortho*-metalation [42–46]. As *ortho*-selective direct arylation promotes a cross-coupling reaction at the α-position of the pyrrole unit, even in the absence of protection of the β-position, the reaction is expected to be useful for the synthesis of π-conjugated polymers with well-defined linkages [47]. Pyrrole is consistently one of the most popular choices of monomer units for preparing π-conjugated polymers [48–51]. Therefore, the synthesis of pyrrole-based π-conjugated polymers via direct arylation would be an extremely useful step forward in the development of improved organic semiconducting materials. We here report the site-selective polycondensation of 1-(2-pyrimidinyl)pyrrole with 2,7-dibromo-9,9-dioctylfluorene via Ru-catalyzed direct arylation (scheme 3). The reaction for the subsequent removal of the directing group from the pyrrole unit of the polymer is also described (scheme 4).

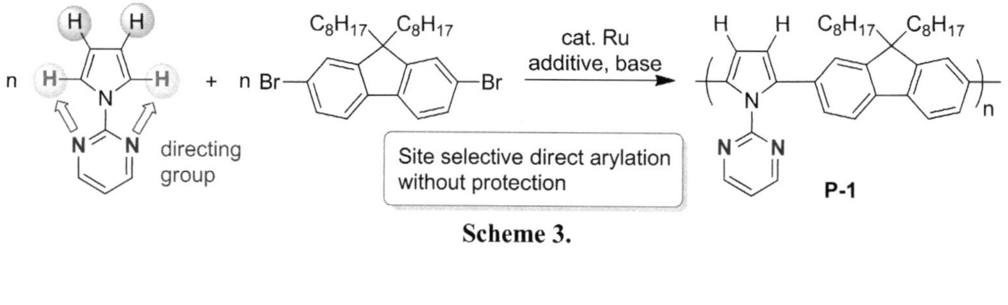

Scheme 3.

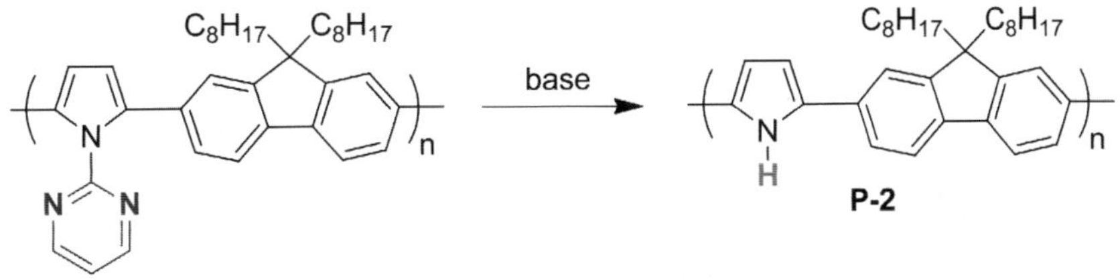

Scheme 4.

2. Experimental

2.1. Typical polycondensation of 1-(2-pyrimidinyl)pyrrole with 2,7-dibromo-9,9-dioctylfluorene via Ru-catalyzed direct arylation: A mixture of $[RuCl_2(p\text{-cymene})]_2$ (15.3 mg, 0.025 mmol), carboxylic acid (0.30 mmol, 60 mol%), K_2CO_3 (415 mg, 3.0 mmol), 2,7-dibromo-9,9-dioctylfluorene (274 mg, 0.50 mmol) and 1-(2-pyrimidinyl)pyrrole (72.6 mg, 0.50 mmol) was stirred in anhydrous *m*-xylene (2.00 mL) for 6-48 h at the indicated temperature under a nitrogen atmosphere. After cooling to room temperature, the solvent was removed by evaporation under vacuum. The residue was poured into an aqueous solution of ethylenediaminetetraacetic acid disodium salt (pH = 8). The suspension was stirred for 1 h at room temperature. The precipitates were separated by filtration and washed with 1 M (mol dm^{-3}) HCl solution, distilled water and MeOH. The filtered residue was poured into hexane and low-molecular-weight fraction was extracted by ultrasonic treatment for 1 h. The precipitates were separated by filtration. If the filtered residue was completely soluble in $CHCl_3$, the residue was dissolved in $CHCl_3$; otherwise, the residue was dissolved in DMF. The solution was filtered through a plug of Celite to remove insoluble materials. The filtrate was concentrated and the polymer (**P-1**) was precipitated by slowly adding the concentrated solution into MeOH.

The polycondensation proceeding under the optimized conditions gave **P-1** in 86% yield (table 1, run 8). GPC (DMF): M_n = 19800, M_w/M_n = 3.82; ^{1}H NMR (600 MHz, DMF-d_7, 293 K): δ 8.91 (2H, d, J = 4.8 Hz, H^i), 7.75 (2H, d, J = 7.8 Hz, H^f), 7.62 (1H, br, H^j), 7.36 (2H, d, J = 7.8 Hz, H^g), 6.88 (2H, br, H^e), 6.61 (2H, br, H^h), 1.63 (4H, br, H^d), 1.30-0.87 (20H, m, H^b), 0.83 (6H, t, J = 7.2 Hz, H^a), 0.28 (4H, br, H^c); $^{13}C\{^1H\}$ NMR (150 MHz, DMF-d_7, 293 K): δ 160.4, 160.0, 151.5, 140.3, 138.6, 133.3, 128.2, 123.1, 121.8, 120.8, 111.4, 55.8, 41.2 (CH_2), 32.7 (CH_2), 30.8 (CH_2), 30.3 (CH_2), 30.1 (CH_2), 24.8 (CH_2), 23.4 (CH_2), 14.6 (CH_3).

2.2 Polymer reaction for the removal of 2-pyrimidinyl substituent on the pyrrole unit of the polymer: A mixture of KOH (1.26 g, 22.5 mmol) and **P-1** (79.8 mg, 0.15 mmol; table 1, run 8) was stirred in anhydrous *N*-methylpyrrolidone (30.0 mL) for 48 h at 200 °C under a nitrogen atmosphere. After cooling to room temperature, the mixture was poured into distilled water. The suspension was stirred for 1 h at room temperature. The precipitates were separated by filtration and washed with distilled water, MeOH and hexane. The filtered residue was dissolved in $CHCl_3$ and the solution was filtered through a plug of Celite to remove insoluble materials. The filtrate was concentrated and dried in a vacuum environment to give an orange solid of poly[(1-*H*-pyrrole-2,5-diyl)-(9,9-dioctylfluorene-2,7-diyl)] (**P-2**) in 82% yield (recovery of polymer). The recovery of polymer was calculated on the basis of the amount of **P-1** taking into account the quantitative reaction. **P-2** is somewhat air sensitive. Since the color of **P-2** was gradually changed from orange to dark blue in air, the polymer was stored under a nitrogen atmosphere. GPC (DMF): M_n = 15500, M_w/M_n = 2.32; ^{1}H NMR (400 MHz, $CDCl_3$, 293 K): δ 8.71 (1H, br, *N*-H), 7.73 (2H, d, J = 7.6 Hz, H^f), 7.57 (2H, d, J = 7.6 Hz, H^g), 7.52 (2H, br, H^e), 6.71 (2H, br, H^h), 2.07 (4H, br, H^d), 1.16 (20H, br, H^b), 0.67 (10H, br, H^a, H^c); $^{13}C\{^1H\}$ NMR (100 MHz, $CDCl_3$, 293 K): δ 151.8, 139.6, 134.0, 131.3, 122.8, 120.2, 118.3, 108.3, 55.4, 40.8 (CH_2), 31.9 (CH_2), 30.2 (CH_2), 29.5 (CH_2), 29.4 (CH_2), 24.0 (CH_2), 22.7 (CH_2), 14.2 (CH_3).

3. Results and Discussion

3.1. Polycondensation

The investigation was conducted employing the polycondensation of 1-(2-pyrimidinyl)pyrrole with 2,7-dibromo-9,9-dioctylfluorene, as the latter possesses good reactivity toward direct arylation polycondensation, and high selectivity against side reactions [25,29–35]. The reaction was firstly examined in the presence of $[RuCl_2(p\text{-cyment})]_2$ (5 mol%), 1-adamantanecarboxylic acid (1-AdCOOH) (60 mol%), and K_2CO_3 (6 equiv) in *m*-xylene according to the literature [40], giving the corresponding polymer (**P-1**) in good yield (table 1, run 6). In contrast, use of the related pyrrole derivatives, 1-methylpyrrole and 1-phenylpyrrole, as monomers did not give the corresponding polymeric products under the same reaction conditions (scheme 5). The control experiments revealed that the presence of the 2-pyrimidinyl substituent induced *ortho*-metalation at the α-position of the pyrrole monomer and promoted the direct arylation.

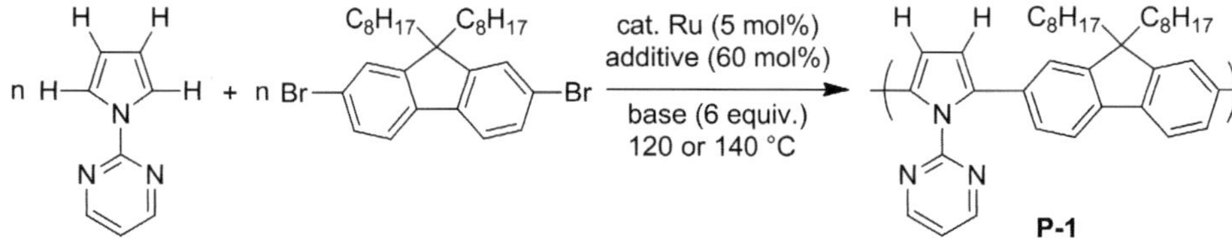

Table 1. Results of polycondensation of 1-(2-pyrimidinyl)pyrrole with 2,7-dibromo-9,9-dioctylfluorene under various conditions[a].

Run	Additive[b]	Temp. / °C	Time / h	Yield[c] / %	GPC (DMF)[d]	
					M_n	M_w/M_n
1	1-AdCOOH	120	24	63	9700	1.73
2	PivOH	120	24	87	11100	1.84
3	MesCOOH	120	24	7	7600	2.00
4	AcOH	120	24	24	7400	1.53
5	PCCA	120	24	53	8300	1.77
6	1-AdCOOH	120	48	87	13100	2.66
7	PivOH	120	48	82[e]	14700	2.35
8	PivOH	140	48	86[e]	19800	3.82
9[f]	-	120	24	20	7600	1.84
10[g]	-	120	24	39	7900	1.40
11[f]	1-AdCOOH[h]	120	24	31	8800	1.95
12[g]	PivOH[h]	120	24	43	11200	2.00

[a] Reactions were conducted using $[RuCl_2(p\text{-cymene})]_2$ (5 mol%), additive (60 mol%), and K_2CO_3 (6 equiv.) in *m*-xylene (2 mL).
[b] MesCOOH : 2,4,6-trimethylbenzoic acid; PCCA: 1-phenyl-1-cyclopentanecarboxylic acid.
[c] Reprecipitation from $CHCl_3$/methanol.
[d] Calibrated by GPC (a DMF solution of LiCl (0.01 M)), polystyrene standards.
[e] Reprecipitation from DMF/methanol.
[f] $[Ru(1\text{-AdCOO})_2(p\text{-cymene})]$ (10 mol%) was used instead of $[RuCl_2(p\text{-cymene})]_2$ (5 mol%).
[g] $[Ru(PivO)_2(p\text{-cymene})]$ (10 mol%) was used instead of $[RuCl_2(p\text{-cymene})]_2$ (5 mol%).
[h] 40 mol%.

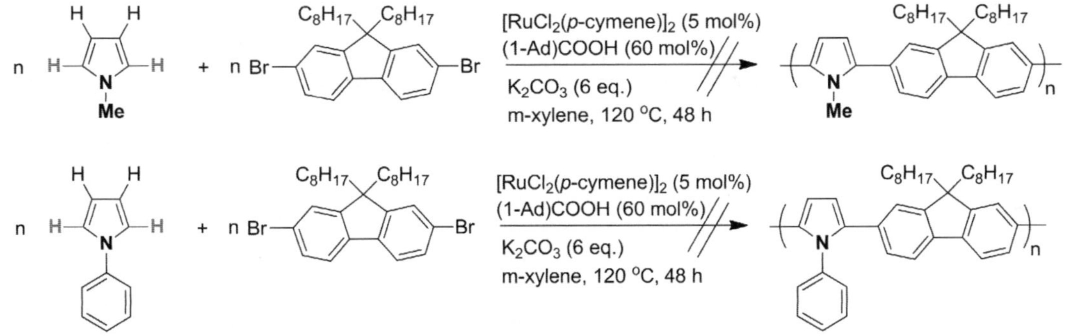

Scheme 5.

In order to determine the appropriate reaction conditions, the reaction was next conducted under a variety of conditions, as summarized in table 1. In terms of the additive in the catalytic system, pivalic acid (PivOH) was found to be superior to the other carboxylic acids tested (runs 1–5). The carboxylic acid serves as a carboxylate ligand under the basic conditions used, which assists the deprotonation of the pyrrole unit in the direct arylation [31–34,52]. As PivOH has the lowest acidity ($pK_a = 5.01$) [53], the high basicity of its conjugated carboxylate ligand is likely to promote a smooth reaction. Increases in reaction time and temperature improved the molecular weight of the products (runs 6–8). In addition, Ru complexes, [Ru(PivO)$_2$(*p*-cymene)] and [Ru(1-AdCOO)$_2$(*p*-cymene)] [54,55], were also prepared and examined as catalysts (runs 9–12), as the use of structurally well-defined Ru complexes was expected to reveal the scope of the direct arylation. Although the use of these complexes was less catalytically competent than that of the *in situ* prepared catalytic system used in the present study, a similar trend was observed with respect to the conjugated carboxylate ligand of the complexes.

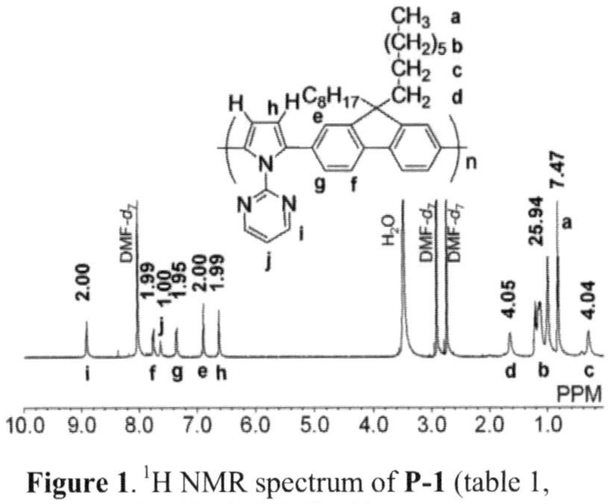

Figure 1. ^{1}H NMR spectrum of **P-1** (table 1, run 8, 600 MHz, in DMF-d_7).

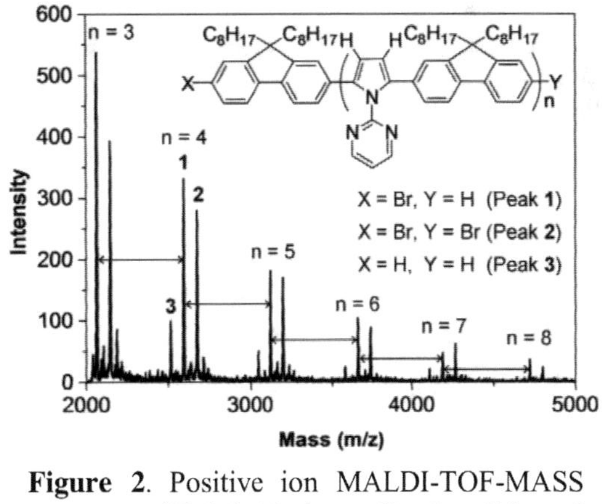

Figure 2. Positive ion MALDI-TOF-MASS spectrum of **P-1** (table 1, run 8) using dithranol as a matrix and des-Arg[1]-Bradykinin, Anglotensin I and Glu[1]-Flibrinopeptide B as external standards.

As the polymeric products were soluble in DMF, the chemical structure of **P-1** was elucidated using NMR and MALDI-TOF-MASS spectroscopy. The ^{1}H NMR spectrum of **P-1** (run 8) is shown in figure 1. All the signals were assignable to the protons in the recurring unit; the signal assignable to the protons at the β-position of the pyrrole unit was observed at δ 6.6 ppm, while that assigned to the proton at the α-position of the pyrrole unit (δ 7.8 ppm) [40] had a negligible intensity. These data indicate that the site-selective direct arylation smoothly proceeded at the α-position of the pyrrole unit, affording well-defined **P-1**. The MALDI-TO-MASS indicates a structure of **P-1** that is consistent with that obtained from the NMR spectrum (figure 2); the main peaks correspond to the recurring unit, and there are no peaks corresponding to the branched structures. In order to elucidate further details of the Ru catalytic system, control experiments were conducted using a Pd catalytic system. When Pd-catalyzed direct arylation of 1-(2-pyrimidinyl)pyrrole with 2,7-dibromo-9,9-dioctylfluorene was carried out under the same reaction conditions as in our previous reports [31,32], the reaction did not give a polymeric product. Alternatively, the reaction of 1-phenylpyrrole with 2,7-dibromo-9,9-dioctylfluorene gave a polymer containing a large insoluble fraction (scheme 6). The IR spectrum of the DMF-insoluble fraction was essentially identical to that of the DMF-soluble fraction, while the MALDI-TOF-MASS spectral analysis suggested the existence of irregular structures. These results indicate that the reaction afforded uncontrollable cross-linked structures at undesired C–H bonds of each aromatic unit [20–22,26–28,29–33,56]. These data also support the occurrence of site-selective polycondensation by Ru-catalyzed direct arylation via *ortho*-metalation, giving the well-defined structure of **P-1**.

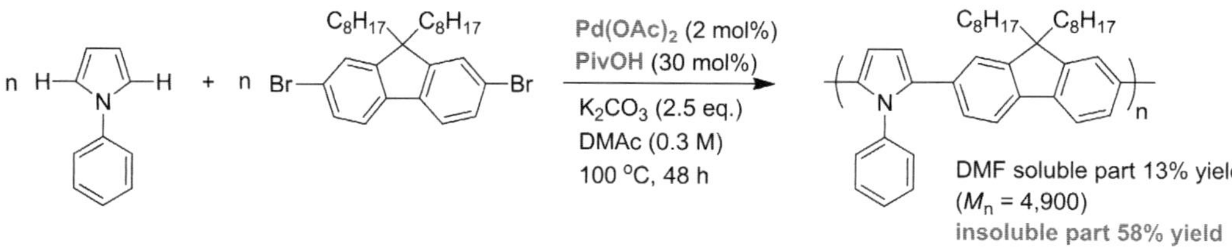

Scheme 6.

3.2. Polymer reaction

In terms of π-conjugated polymers, steric hindrance due to bulky substituents on the aromatic recurring units sometimes leads to limited π-conjugation along the polymer main chain [33,57,58]. There are a number of reports concerning reactions used to remove directing groups [40,41,59]; therefore, we utilized this knowledge to remove the 2-pyrimidinyl substituent on the pyrrole unit of **P-1** (scheme 7). The 2-pyrimidinyl group was efficiently cleaved from **P-1** by treatment with KOH, giving the corresponding polymer (**P-2**) with high recovery. The GPC data of **P-2** indicate that decomposition of the polymer main chain is likely to be negligible owing to the thermal stability.

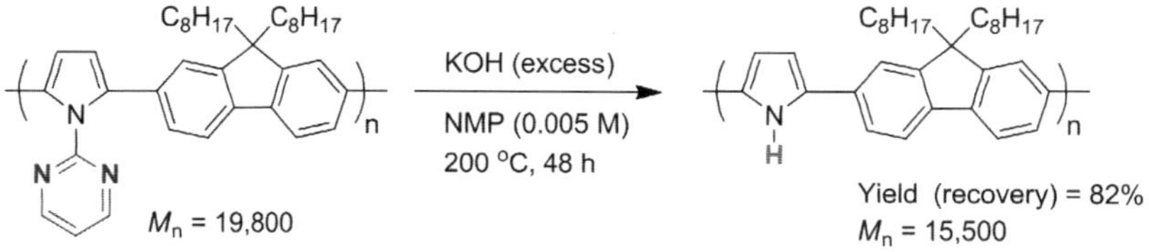

Scheme 7.

The ^{1}H NMR spectrum of **P-2** reveals efficient conversion of the pyrrole unit, as shown in figure 3. The signals assigned to the 2-pyrimidinyl substituent that are evident in the spectrum of **P-1** (figure 1) are no longer visible, but a new signal assignable to the 1-H proton of the pyrrole unit can be observed at δ 8.7 ppm, which was confirmed by the decrease in signal upon addition of D_2O. The integral ratios of the signals are essentially identical to the corresponding recurring unit of **P-2**. The $^{13}C\{^1H\}$ NMR and MALDI-TOFMASS spectra provide further evidence of efficient removal.

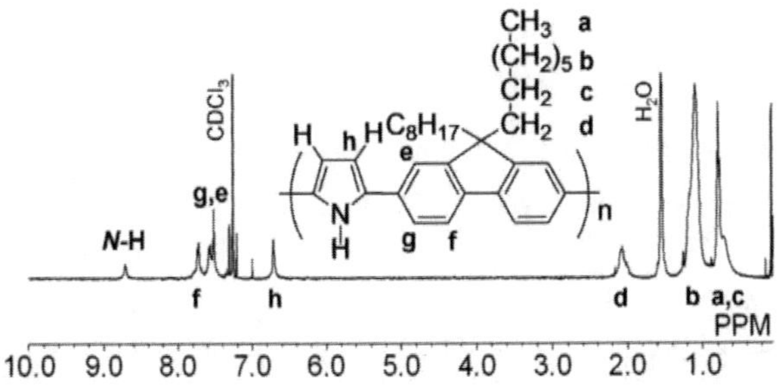

Figure 3. ^{1}H NMR spectrum of **P-2** (400 MHz, in $CDCl_3$).

3.3. Optical properties

The optical properties of the polymers were compared using UV-vis absorption spectroscopy. Figure 4 shows the UV-vis absorption spectra of the polymers in thin film state. **P-1** gave similar absorptions in the solution and thin-film states, which suggests weak interchain interactions such as π–π stacking in the film state. This is likely caused by the steric hindrance from the bulky 2-pyrimidinyl substituent. The spectrum of **P-2** exhibits a red shift of the absorption by 26 and 41 nm in solution and thin film states, respectively, when compared with **P-1**. Removal of the 2-pyrimidinyl substituent led to the elongation of π-conjugation along the main chain of polymer, resulting in the red shift of absorption. In terms of the emission properties, the solution and film of **P-1** exhibited cyan (452 nm) and green (483 nm) emission, respectively; whereas, **P-2** showed only weak emission (498 nm) in solution.

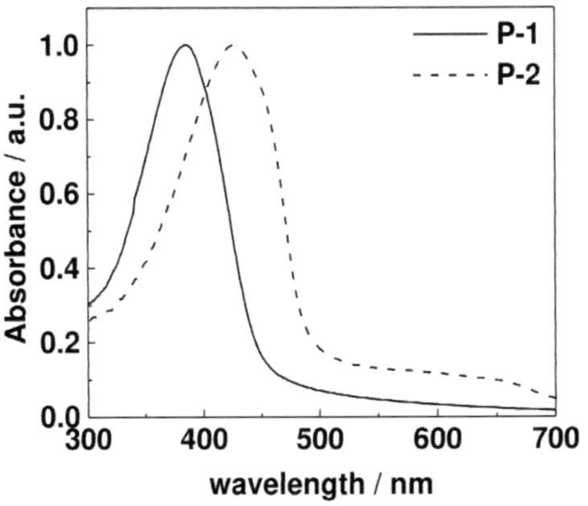

Figure 4. UV-vis absorption spectra of **P-1** and **P-2** in film.

4. Conclusion

Site-selective polycondensation was demonstrated using Ru-catalyzed direct arylation via *ortho*-metalation. The molecular design of the aromatic monomer bearing the directing group efficiently suppressed the formation of branched and cross-linked structures during the direct arylation polycondensation. Subsequent removal of the 2-pyrimidinyl directing group enhanced the coplanarity of the π-conjugated polymer. Although this polycondensation reaction has limited scope at present, and the introduction and removal of the directing group is not ideal with respect to atom efficiency, this strategy should present new possibilities for direct arylation polycondensation in the synthesis of π-conjugated polymers with well-defined linkages between the recurring units.

References

[1] Bernius M T, Inbasekaran M, ÓBrien J and Wu W S 2000 *Adv. Mater.* **12** 1737-50
[2] Grimsdale A C, Chan K L, Martin R E, Jokisz P G and Holmes A B 2009 *Chem. Rev.* **109** 897-1091
[3] Huang F, Wu H and Cao Y 2010 *Chem. Soc. Rev.* **39** 2500-21
[4] Boudreault P-L T, Najari A and Leclerc M 2011 *Chem. Mater.* **23** 456-69
[5] Chen Y-J, Yang S-H and Hsu C-S 2009 *Chem. Rev.* **109** 5868-923
[6] Yamamoto T 2010 *Bull. Chem. Soc. Jpn.* **83** 431-55
[7] Sakamoto J, Rehahn M, Wegner G and Schlüter A D 2009 *Macromol. Rapid Commun.* **30** 653-87
[8] Carsten B, He F, Son H J, Xu T and Yu L 2011 *Chem. Rev.* **111** 1493-528
[9] Okamoto K and Luscombe C K 2011 *Polym. Chem.* **2** 2424-34
[10] Roncali J 2007 *Macromol. Rapid Commun.* **28** 1761-75
[11] Yokozawa T, Nanashima Y and Ohta Y 2012 *ACS Macro Lett.* **1** 862-6
[12] Cheng Y-J and Luh T-Y 2004 *J. Organomet.Chem.* **689** 4137-48
[13] Facchetti A, Vaccaro L and Marrocchi A 2012 *Angew. Chem., Int. Ed.* **51** 3520-3
[14] Wang Q, Takita R, Kikuzaki Y and Ozawa F 2010 *J. Am. Chem. Soc.* **132** 11420-1
[15] Wang Q, Wakioka M and Ozawa F 2012 *Macromol. Rapid Commun.* **33** 1203-7
[16] Kumar A and Kumar A 2010 *Polym. Chem.* **1** 286-8
[17] Berrouard P, Najari A, Pron A, Gendron D, Morin P-O, Pouliot J-R, Veilleux J and Leclerc M 2012 *Angew. Chem. Int. Ed.* **51** 2068-71
[18] Beaupré S, Pron A, Drouin S H, Najari A, Mercier L G, Robitaille A and Leclerc M 2012 *Macromolecules* **45** 6906-14
[19] Berrouard P, Dufresne S, Pron A, Veilleux J and Leclerc M 2012 *J. Org. Chem.* **77** 8167-73
[20] Kowalski S, Allard S and Scherf U 2012 *ACS Macro Lett.* **1** 465-8
[21] Rudenko A E, Wiley C A, Stone S M, Tannaci J F and Thompson B C 2012 *J. Polym. Sci. Part A: Polym. Chem.* **50** 3691-7
[22] Chang S-W, Waters H, Kettle J, Kuo Z-R, Li C-H, Yu C-Y and Horie M 2012 *Macromol. Rapid Commun.* **33** 1927-32
[23] Allard N, Najari A, Pouliot J-R, Pron A, Grenier F and Leclerc M 2012 *Polym. Chem.* **3** 2875-9
[24] Abdo N I, El-Shehawy A A, El-Barbary A A and Lee J-S 2012 *Eur. J. Org. Chem.* **2012** 5540-51
[25] Wakioka M, Kitano Y and Ozawa F 2013 *Macromolecules* **46** 370-4
[26] Guo Q, Dong J, Wan D, Wu D and You J 2013 *Macromol. Rapid Commun.* **34** 522-7
[27] Mercier L G, Aïch B R, Najari A, Beaupré S, Berrouard P, Pron A, Robitaille A, Tao Y and Leclerc M 2013 *Polym. Chem.* **4** 5252-60
[28] Rudenko A E, Wiley C A, Tannaci J F and Thompson B C 2013 *J. Polym. Sci. Part A: Polym. Chem.* **51** 2660-8
[29] Lu W, Kuwabara J and Kanbara T 2011 *Macromolecules* **44** 1252-5
[30] Lu W, Kuwabara J, Iijima T, Higashimura H, Hayashi H and Kanbara T 2012 *Macromolecules* **45** 4128-33
[31] Lu W, Kuwabara J and Kanbara T 2012 *Polym. Chem.* **3** 3217-9
[32] Fujinami Y, Kuwabara J, Lu W, Hayashi H and Kanbara T 2012 *ACS Macro Lett.* **1** 67-70
[33] Kuwabara J, Nohara Y, Choi S J, Fujinami Y, Lu W, Yoshimura K, Oguma J, Suenobu K and Kanbara T 2013 *Polym. Chem.* **4** 947-53
[34] Yamazaki K, Kuwabara J and Kanbara T 2013 *Macromol. Rapid Commun.* **34** 69-73
[35] Choi S J, Kuwabara J, Kanbara T 2013 *ACS Sustainable Chem. Eng.* **1**, 878-83
[36] Tamba S, Mitsuda S, Tanaka F, Sugie A and Mori A 2012 *Organometallics* **31** 2263-7
[37] Tamba S, Shono K, Sugie A and Mori A 2011 *J. Am. Chem. Soc.* **133** 9700-3
[38] Tamba S, Tanaka S, Okubo Y, Meguro H, Okamoto S and Mori A 2011 *Chem. Lett.* **40** 398-9
[39] Tanaka S, Tanaka D, Tatsuta G, Murakami K, Tamba S, Sugie A and Mori A 2013 *Chem. Eur. J.* **19** 1658-65
[40] Ackermann L and Lygin A V 2011 *Org. Lett.* **13** 3332-5

[41] Ackermann L and Lygin A V 2012 *Org. Lett.* **14** 764-7
[42] Kakiuchi F and Chatani N 2003 *Adv. Synth. Catal.* **345** 1077-101
[43] Alberico D, Scott M E and Lautens M 2007 *Chem. Rev.* **107** 174-238
[44] Ackermann L, Vicente R and Kapdi A R 2009 *Angew. Chem. Int. Ed.* **48** 9792-826
[45] Ackermann L 2011 *Chem. Rev.* **111** 1315-45
[46] Liu C, Zhang H, Shi W and Lei A 2011 *Chem. Rev.* **111** 1780-824
[47] Kawamitsu S and Tomita I 2012 *The 92nd Annual Meeting of The Chemical Society of Japan (Yokohama, Japan, 25-28 March 2012)* 1J2-42
[48] Dhanabalan A, van Dongen J L J, van Duren J K J, Janssen H M, van Hal P A and Janssen R A J 2001 *Macromolecules* **34** 2495-501
[49] Lan L, Peng J, Sun M, Zhou J, Zou J, Wang J and Cao Y 2009 *Org. Electron.* **10** 346-51
[50] Sun M, Wang L and Yang W 2010 *J. Appl. Polym. Sci.* **118** 1462-8
[51] Kitamura C, Tanaka S and Yamashita Y 1996 *Chem. Mater.* **8** 570-8
[52] Lafrance M and Fagnou K 2006 *J. Am. Chem. Soc.* **128** 16496-7
[53] The pK_a values of the carboxylic acids are as follow, pivalic acid (5.01), 1-adamantanecarboxylic acid (4.86), acetic acid (4.76), 1-phenyl-1-cyclopentanecarboxylic acid (4.39), 2,4,6-trimethylbenzoic acid (3.85). a) pK_a data compiled by Williams R http://research.chem.psu.edu/brpgroup/pK_a_compilation.pdf b) http://www.guidechem.com/msds/77-55-4.html c) http://www.guidechem.com/msds/480-63-7.html (accessed March 2013)
[54] Ackermann L, Novák P, Vicente R and Hofmann N 2009 *Angew. Chem. Int. Ed.* **48** 6045-8
[55] Arockiam P B, Fischmeister C, Bruneau C and Dixneuf P H 2010 *Angew. Chem. Int. Ed.* **49** 6629-32
[56] Zhao L, Bruneau C and Doucet H 2013 *ChemCatChem* **5** 255-62
[57] Andersson M R, Thomas O, Mammo W, Svensson M, Theanderc M and Inganäsc O 1999 *J. Mater. Chem.* **9** 1933-40
[58] Kim J-B and Lim S-T 1996 *Polym. Bull.* **37** 297-303
[59] Nishino M, Hirano K, Satoh T and Miura M 2012 *Angew. Chem. Int. Ed.* **51** 6993-7

Asymmetric polymerisation in liquid crystals and resultant electro-chiroptical effect: Structure organising polymerisation and chiral charge carrier "chiralion"

Hiromasa Goto

Division of Materials Science, Faculty of Pure and Applied Sciences, University of Tsukuba, Tsukuba, Ibaraki 305-8573, Japan

E-mail: gotoh@ims.tsukuba.ac.jp

Abstract. Electrochemical synthesis in liquid crystal (LC) affords conducting polymers having LC molecular order and electro-activity. The polymerisation method can be referred to as structure organising polymerisation (SOP). The optical textures of the polymers thus prepared appear very similar to that of the LC electrolyte solution used for the polymerisation. Especially, polymers prepared in cholesteric LC (chiral LC) having structural chirality show doping-dedoping (redox) driven change in chiroptical activity (controllable circular dichroism and optical rotation), as "electro-chiroptical effect". The polymer films exhibit interference colour and electrochemically driven refractive index modulations. The chiroptical activity of the polymer prepared in cholesteric LC comes from axial chirality of the helical structure.

1. Introduction

Chiral liquid crystals are fascinating materials due to their liquid-like fluidity, crystal-like order, and hierarchic structure [1]. The helical aggregation of rod-like molecules in cholesteric liquid crystal (cholesteric LC) results in periodicity, and the three-dimensional (3D) molecular arrangement produces structural chirality. Many living organisms possess cholesteric LC structures [2–5]. The exoskeleton of insects with cholesteric order shows selective reflection of light. These are referred to as photonic insects. For example, the exoskeleton of golden beetles has cholesteric order at the microscopic level. A jewel beetle cuticle consists of the alternation of multi-layers having liquid crystal (LC) order [6]. Life forms contain structures related to LCs, e.g., brain, cells, and muscle. Furthermore, DNA shows LC structure at certain concentration in water [7].

The helical arrangement of a cholesteric LC on a macromolecular scale greatly enhances chiroptical properties. The optical rotation of a chiral molecule can be enhanced in the cholesteric LC phase. Therefore, chiral conjugated polymers with cholesteric LC-like macromolecular formations are expected to exhibit optical activity.

LCs can be useful reaction medium, having both liquid-like fluidity and ordered structures that can be imprinted on the forming molecules. A LC can thus function as both a reaction medium and a template. Chemical reactions in cholesteric LCs are particularly effective in producing chiral compounds from achiral materials, since the structural chirality of the cholesteric LC provides a three-dimensional (3D) sequential chiral matrix for the chemical reaction. Polymerisation in cholesteric liquid crystals performs "matrix imprinting" (MAT-I) for the resultant polymer.

The author synthesised mono-subsitituted polyacetylene derivatives having chiral groups in order to obtain ferroelectric liquid crystalline conjugated polymers based on the original idea of the author in Shirakawa lab (U. Tsukuba). The chiral chemistry and S_N2-type Mitsunobu reaction have been introduced in the lab through the idea and experiments in 1994. The polymer thus synthesised showed chiroptical activity and beautiful spiral patterns derived from the helical structure with chirality [8,9]. Here, in this lab, synthesis of polyacetylene in nematic LC has been carried out as the previous collaborative study by Araya and Shirakawa et al. [10].

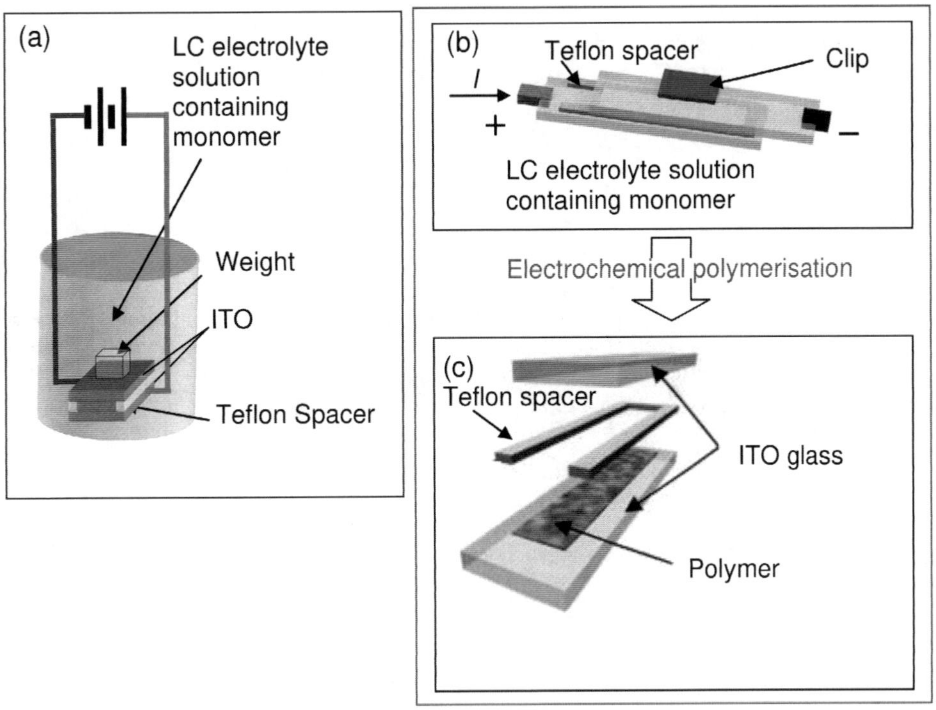

Figure 1. (a) Prototype electrochemical polymerisation in liquid crystal (LC). (b) Sandwich cell LC polymerisation. (c) Detachment of the cell after the polymerisation. ITO = indium tin oxide coated glass. Reproduced by permission of the Royal Society of Chemistry from ref [21] (Figure 1(d,e)).

After several months (November 1994) from the synthesis of the chiral liquid crystal polyacetylene derivative showing spiral structure, synthesis of polyacetylene by using the chiral smectic C (SmC*) for the reaction field was proposed in the lab based on the results obtained by the author. However, at the time, the author mentioned that cholesteric LC (chiral nematic LC, N*-LC) should be employed in place of SmC* LC materials because the synthesis of polyacetylene in the SmC* LC appeared to be difficult. Acetylene gas cannot permeate into the SmC* containing the catalyst due to its high viscosity, resulting in low polymerisation activity at the SmC* reaction field, and synthesis of a large volume of SmC* with multi-step process is required for it. The author suggested the use of an additional chiral dopant (chiral inducer) to nematic liquid crystals for the production of cholesteric LC based on the original idea of the author. At that time, the lab had no concept of the chiral dopants, whereas electro-dopants, such as electron acceptors and donors, for conjugated polymers had been well studied. Therefore, this was the first idea and example of the introduction of the cholesteric LC reaction matrix for production of conjugated polymers (November, 1994). Later, the helical polyacetylene was synthesised with the development of a new chiral inducer [11–14]. In 2000, the synthesis of polypyrrole with a prototype electrochemical polymerisation in LC was carried out in a vial by the author [14,15]. After much effort, the submerged sandwich method was developed. A sandwiched indium tin oxide (ITO) coated glass electrodes were submerged in a LC electrolyte solution containing pyrrole as a monomer, tetrabutylammonium perchlorate (TBAP) as a supporting electrolyte, and a chiral dopant (chiral inducer), as shown in Figure 1(a) [14]. The polymerisation was carried out by the applied voltage of 3 V at room temperature. The resultant film displayed fingerprint texture very similar to that of the LC electrolyte solution under scanning electron microscopy (SEM) observations. In 2004, a convenient two-sandwich electrochemical polymerisation method using minimal amounts of LC electrolyte solution with good production efficiency was developed (Figure 1(b,c)) as the original idea of the author, who was assistant professor at the time [15]. This result was published in 2002 as a first report of electrochemical polymerisation in chiral LCs [16]. The film thus prepared shows good electro-activity. Furthermore, chiroptical activity (circular dichroism (CD) and optical rotation) of the film could be tuned by an electrochemical redox process. The redox process is equivalent to electrochemical doping (oxidation) and dedoping (reduction). In other words, the resultant polymer films, as expected for an organic semiconductor, show doping/dedoping-driven electro-chiroptical effect. The electro-controllable optical rotation effect is compared to Faraday rotation and Kerr effect. The chiroptical activity is precisely and rapidly controlled with voltages less than 1 V in these materials. These original findings were obtained by the author on 27 July, 2004. This result and method had not been accepted in the lab, while repeated experiments confirmed the effectiveness of this method and phenomenon. Therefore, the author submitted these findings [17,18].

Electrochemical polymerisation in LC affords electroactive polymers having similar structure to that of the electrolyte solution due to transcription of structural chirality from LC to the resultant polymers. However, the surface structure depends on molecular shape, the affinity to the LC electrolyte solution, and the polymerisation activity of the monomer. Also, crystallinity and film-forming properties are related to the morphology of the resultant polymers. In this paper, the term of "chiral inducer" is employed in place of chiral dopant to avoid confusion with electro-dopants (donors and acceptors). The polymerisation of various monomers in LC and resultant properties are discussed.

2. General method of electrochemical polymerisation in LC

The cholesteric electrolyte solution used as the reaction matrix is prepared by adding a chiral inducer [19] (i.e., cholesteryl pelargonate) to 4-*n*-pentyl-4'-cyanobiphenyl (5CB) or 4-*n*-hexyl-4'-cyanobiphenyl (6CB). A nematic LC electrolyte solution is prepared by the same manner but without adding a chiral inducer. Smectic A (SmA) LC having the layer structure of the electrolyte solution is prepared in a mixed LC system (4-*n*-octyloxy-4'-cyanobiphenyl (8OCB) and 6CB).

Prior to electrochemical polymerisation, the LC electrolyte solution containing monomer (bithiophene, ethylene dioxythiophene, pyrrole, furan, or their derivatives) was heated in a vial to ca. 80 °C under an argon atmosphere in order to completely dissolve the constituents in the LC solvent. The LC mixture was injected between two sandwiched ITO coated glass electrodes with a Teflon sheet (thickness of ca. 0.2 mm) as a spacer by a capillary technique. The reaction cell was initially heated to ca. 40 °C, and then gradually cooled to the LC temperature range. A voltage of 4.0 V was then applied across the cell. After 30 min, an insoluble and infusible polymer thin film was obtained at the sandwiched ITO-coated glass electrode of anode side. Then, the sandwich cell was detached. The polymer film on the ITO was washed with solvents (tetrahydrofuran, hexane, and acetone) and dried.

3. Surface structure

Scanning electron microscopy (SEM) image of poly(2,7-di(2-furyl)fluorene) (PFFLF*, * = chiral) thus prepared in the cholesteric LC shows the labyrinth-like concavo-convex structure (Figure 2(a)), although it should be noted that the molecular structure of the polymer differs from that of the cholesteric LC employed as the electrolyte solution [20,21]. The polymer contains no LC molecules due to phase separation in the polymerisation process.

The laser transmission through the polymer film produces a circular Fourier-transformed interference with distinct separation of red, and green laser light due to a random grating formed by the single-pitch helix-derived stripes of the polymer (Figure 2(b)). Poly(3,4-ethylenedioxythiophene) (PEDOT*) from *ter*EDOT prepared in the cholesteric LC appears to display a vortex pattern [22].

A synthetic scheme for the PEDOT* and a polarising optical microscopy (POM) image of the cholesteric electrolyte solution containing the *ter*EDOT as a monomer are shown in Figure 3(a). The SEM images of the resultant polymer confirm that the polymer imprinted the cholesteric LC structure (vortex structure, Figure 3(b)). The magnification image indicates that the vortex consists of fine lamellas. The optical textures originate from individual molecular ordering of the cholesteric LC matrix. The surface morphologies depend on polymerisation activity of the monomers and the crystallinity of the resultant polymers.

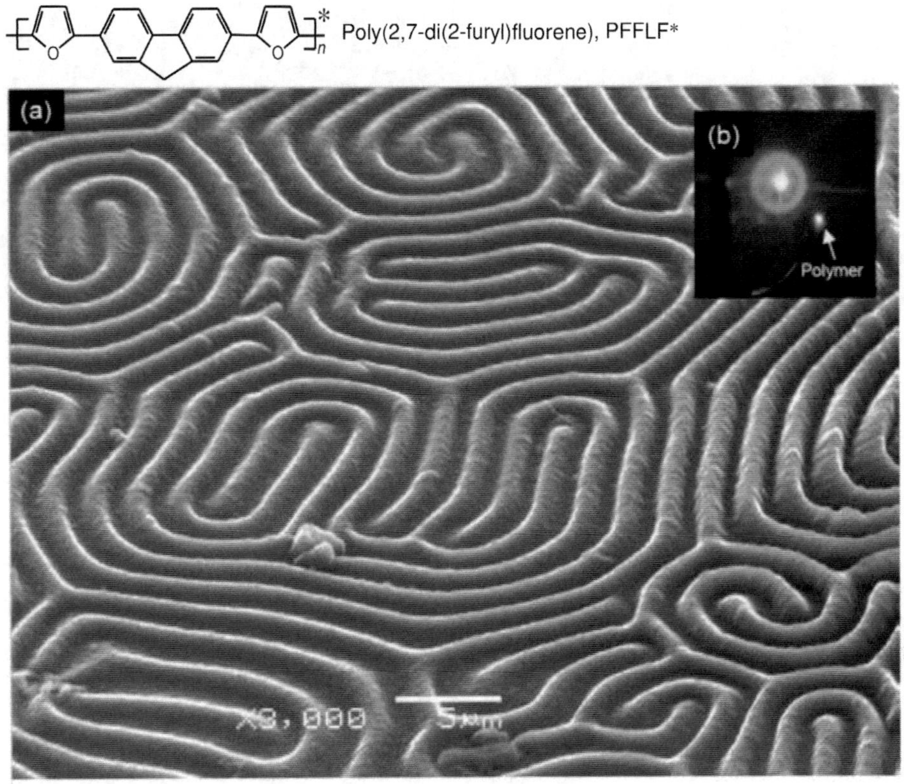

Figure 2. (a) Scanning electron mictroscopy (SEM) image of chiral poly(2,7-di(2-furyl)fluorene) (PFFLF*, * = chiral). (b) Diffraction pattern obtained upon irradiation by red and green laser incident light. Reproduced by permission of the Royal Society of Chemistry (RSC) from ref. [20] (Figure 5(d)).

4. Electrochemically driven change in chiroptical activity

The polymers thus prepared in cholesteric LC exhibit Cotton effect associated with the π–π* transition of the main chain [23]. The circular dichroism (CD) spectra of the reduced (dedoped) polymers reveal a split-type Cotton effect, such as exciton coupling. In the case of the polybithiophene (PBTh*) prepared in cholesteric LC, the CD spectra for the PBTh* film display a positive Cotton effect at 600 nm in the oxidised state. The CD signals cannot be attributed to the cholesteric LC inducer employed in the polymerisation because the weak Cotton effect of the cholesteric LC inducer (cholesteryl pelargonate) is only observed at wavelengths in the range of 240–340 nm. Optical activity of the PBTh* can be changed by adjusting the conditions of the electrochemical redox (electrochemical doping-dedoping) process without changes in optical texture or surface structure. The PBTh* prepared in the cholesteric LC is, therefore, electro-chiroptically active and the CD can be controlled via an electrochemical process. This phenomenon can be referred to as electrochemically driven chiroptical effect (electro-chiroptical effect). Furthermore, control of optical rotation of the polymer is also possible by the redox process. This electrochemical redox state-induced change in optical rotation can be reasonably explained as being due to electrochemical doping/dedoping of the main chain.

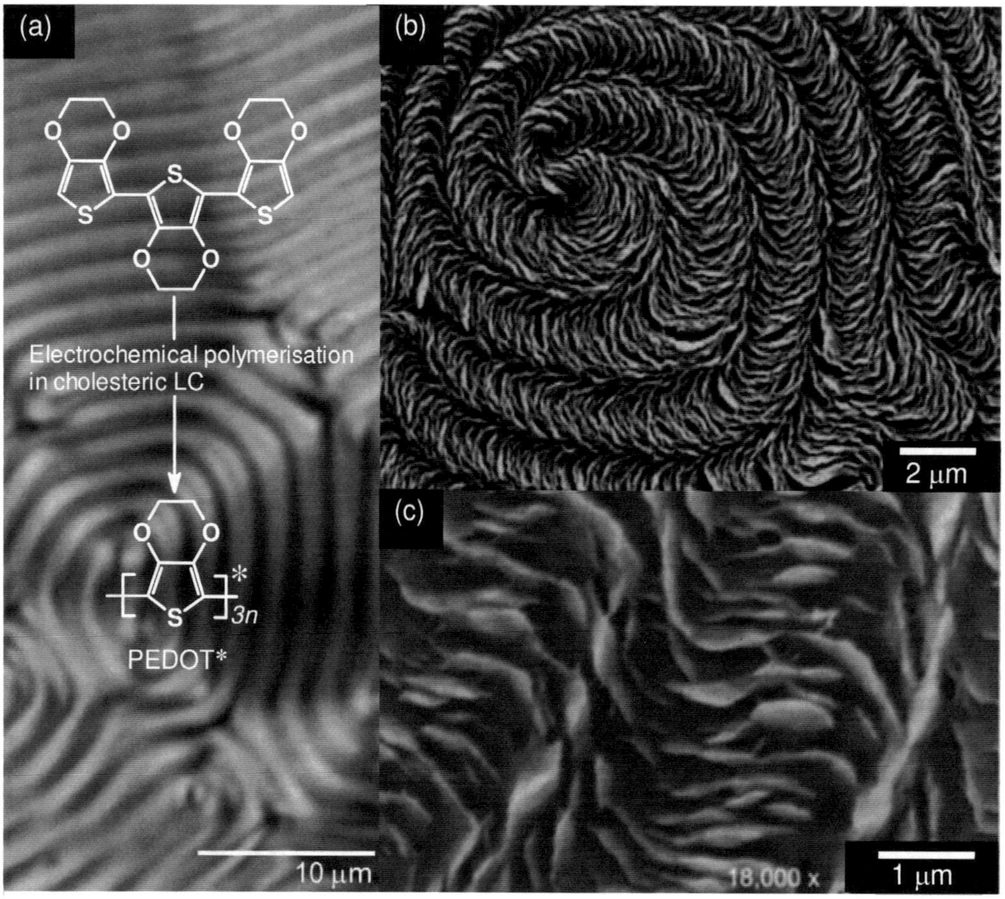

Figure 3. (a) Polarising optical microscopy (POM) image of cholesteric liquid crystal (cholesteric LC) electrolyte solution containing monomer and a scheme for preparation of chiral poly(3,4-ethylenedioxythiophene) (PEDOT*). (b) SEM image of the PEDOT* prepared in cholesteric LC electrolyte solution. (c) Magnification image. Reproduced by permission of the Royal Society of Chemistry (RSC) from ref. [21] (Figure 3(a)).

5. Helical polarons – chiralion

Electron spin resonance (ESR) spectroscopy measurements during the redox process confirm the generation of charge carriers (polarons, radical cations) by electrochemical doping. Increase of applied voltage for the polymer progresses electrochemical doping by perchlorate ion in the electrolyte solution. Figure 4 shows ESR line shapes of the PBTh* prepared in cholesteric LC. The intensity of the ESR signal (I_{ESR}) was increased and line width decreased with applied voltage up to 0.4 V because of increase of polarons unit in the main chain. The delocalisation of the polarons along the main chain resulted narrowing of the signals. However, the I_{ESR} was decreased and the signal width was broadened due to generation of bipolarons (dications) by further application of higher voltage.

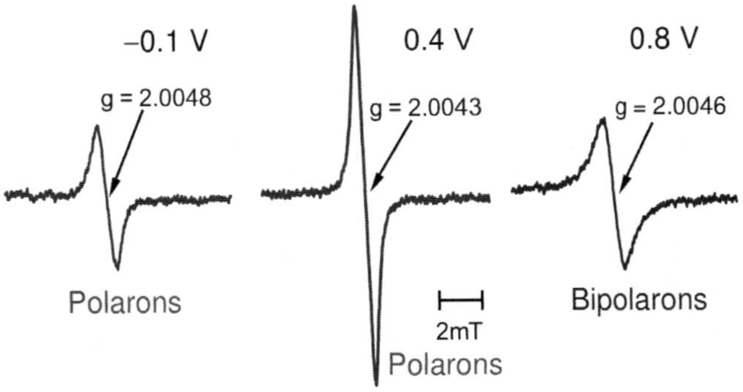

Figure 4. ESR line shapes of the chiral polybithiophene (PBTh*) prepared in cholesteric LC at −0.1 V, 0.4 V, and 0.8 V vs. Fc/Fc$^+$.

Figure 5 summarises a set of the ESR results. The g-value of the polymer decreased with increase of polarons accompanied by increase of applied voltage. Subsequent application of higher voltage allows increase of the g-value (Figure 5(a)). The I_{ESR} increased with applied voltage up to 0.4 V because of increase of the polarons unit in the main chain. Further applied voltage, as a progress of the electrochemical doping, decreased the I_{ESR} due to the generation of bipolarons (Figure 5(b)). Spin (radical) number (ρ) of the polymer once increased, and decreased gradually with the voltage. The maximum of the ρ corresponds to an inflection point of the I_{ESR}. The change from polarons to bipolarons is accompanied by change in magnetic properties of the doped polymer from paramagnetism (polarons) to diamagnetism (bipolarons).

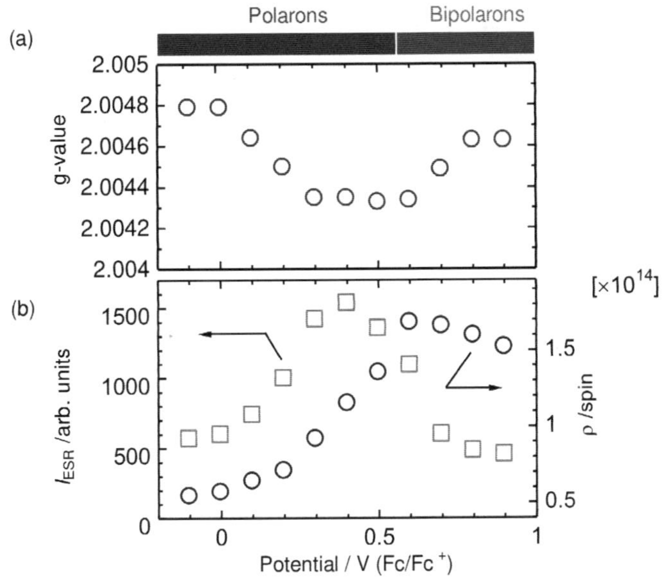

Figure 5. ESR results for the PBTh*. (a) g-value vs. *E*. (b) I_{ESR} and ρ vs. *E*.

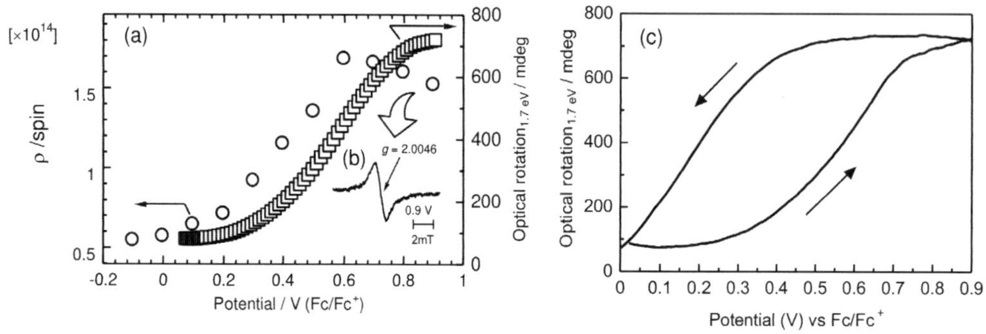

Figure 6. (a) ESR results and optical rotation for the PBTh*. Circle: spin (radical) number of the polymer vs. potentials (*E* vs Fc/Fc^+). Square: optical rotation values plotted at 1.7 eV (729 nm) during the electrochemical doping process. (b) ESR line shape at 0.4 V vs. Fc/Fc^+. (c) Optical rotation hysteresis effect for the PBTh* as a function of potential vs. Fc/Fc^+ reference in 0.1 M tetrabutyl ammonium perchlorate (TBAP)/acetonitrile solution. Reproduced by permission of Elsevier Limited (Figure 6(a,b)), Elsevier, from ref. [24] (Figure 3).

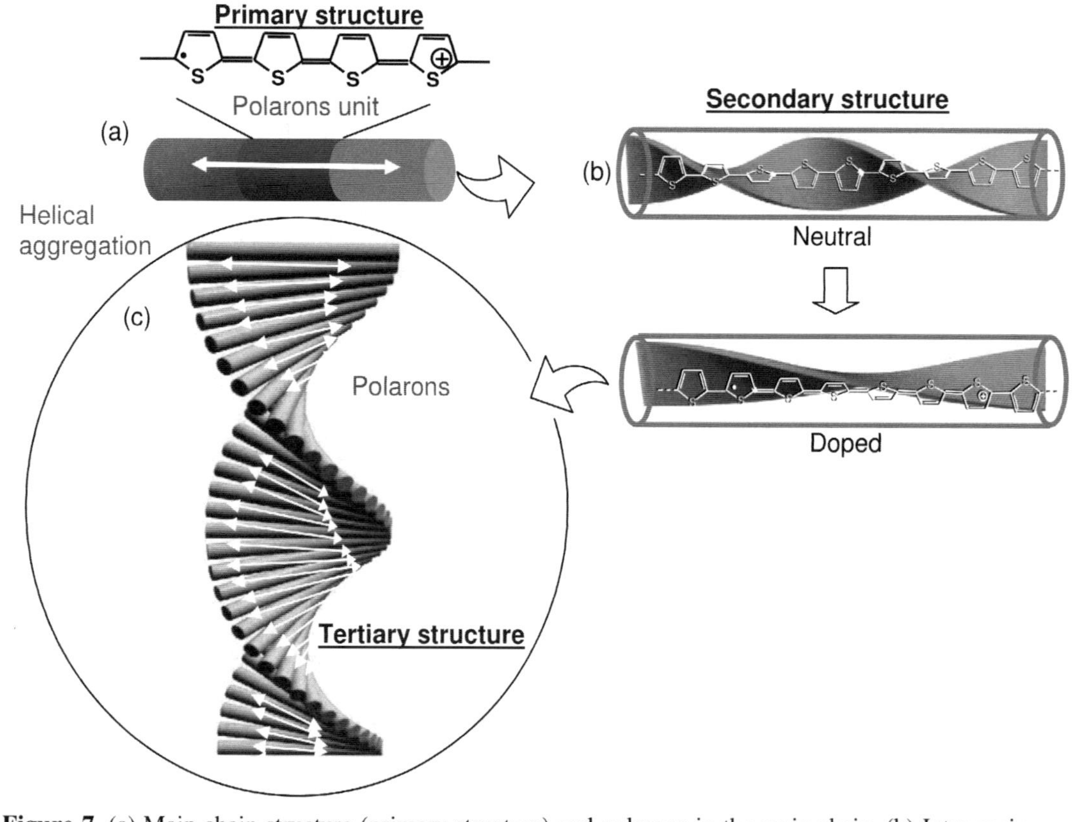

Figure 7. (a) Main chain structure (primary structure) and polarons in the main chain. (b) Intra-main-chain helical twisting and change in the helicity by doping (secondary structure). (c) Chiral aggregation structure (tertiary structure) to show the fingerprint texture observable under POM and SEM.

Figure 6(a) shows change in optical rotation and ρ of the polymer with applied voltages vs. Fc/Fc^+ as a reference. Increase of the optical rotation and ρ value were observed with increasing of the voltage. The results indicate that polarons on the main chain are in a three-dimensional (3D) chiral environment. The increase of the optical rotation values corresponds to increase of the spins with doping. Optical rotation hysteresis effect was also observed, as shown in Figure 6(c). This demonstrates that the optical rotation is controlled by doping. Polarons distribute on the helically formed main chains at appropriate potentials [24]. Furthermore, introduction of quinonoid structure (polarons unit) of the helical main chain tunes chiroptical activity by the electrochemical doping (Figure 7(b)). Also, intercalation of the dopant between the individual main chains tunes the chiral aggregation. The inter-chains electronic communication is depressed in the doped state due to extension of the inter-chain distance by the intercalation. The intra-chain (generation of polarons in the main chain) and inter-chain (intercalation between main chains) electrochemical doping-dedoping (redox process) can control the chiroptical properties. The polarons band of the polymers shows the CD signals, indicating the polarons as charge carrier have chioptical activity. The polarons with chirality is defined as "chiralion".

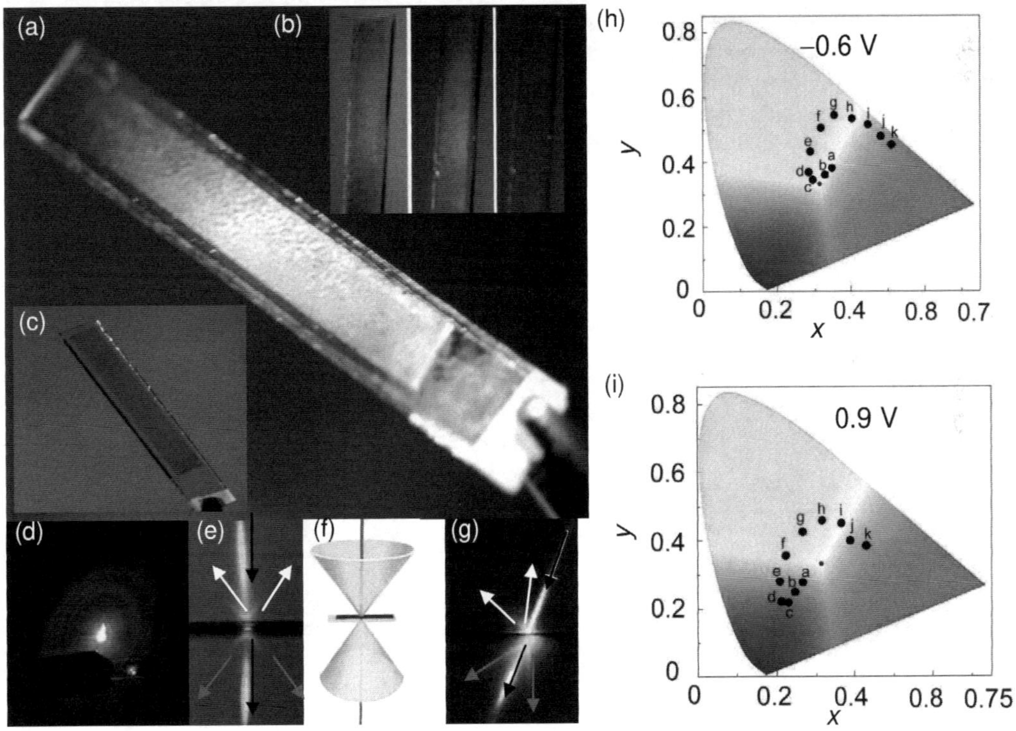

Figure 8. [(a) and (b)] Polymer film on ITO under white light at various reflection angles. (c) Natural appearance of polymer film. (d) Diffraction pattern under irradiation with combined red, green, and blue laser light. (e) Green laser diffraction at normal incidence. Black arrow indicates incident direction, white arrows denote reflection directions, and blue arrows denote transmission directions. (f) Schematic of three-dimensional hollow cone drawn by the transmission and reflection of light at normal incidence. (g) Green laser reflection and transmission at θ = 20°. (h) Transmission colours in CIE colour chromaticity space calculated from transmission spectra for normally incident light at –0.6 V (h) and +0.9 V (i) vs. Fc/Fc^+. Detection angles (θ): a) 24°, b) 26°, c) 28°, d) 30°, e) 32°, f) 34°, g) 36°, h) 38°, i) 40°, j) 42°, and k) 44°. Reprinted with permission from *J. Appl. Phys.* 105, 114906 (2009) [25] (Figures 2,3). Copyright [2009], AIP Publishing LLC.

Possible helical aggregation and distribution of polarons along the main chain are displayed in Figure 7. The polymer forms hierarchical chiral structures. The main chain consists of polythiophene with polarons unit in the doped state (**primary structure**), Figure 7(a). The main chain is twisted. The twist structure is produced by the matrix imprinting (MAT-I) from cholesteric LC during the polymerisation (**secondary structure**, Figure 7(b)). The main chain aggregates to form helical steps like structure as a chiral aggregation form (**tertiary structure**, Figure 7(c)).

6. Variable interference colour modulation

The PBTh* reflects rainbow colour upon incident light due to periodic dielectric structure produced by transcription of structural chirality from the cholesteric LC electrolyte solution. The colour of the reflected light varies with the detection angle, indicating that the molecular ordering of the present polymer acts as a reflection grating. Furthermore, the natural colour of the polymer changes from dark red to dark blue according to the oxidation, while the structural colour (interference reflection colour derived from its periodic structure) changes from light blue to orange under obliquely incident white light at a certain angle. The interference colour can be tuned by the electrochemical doping-dedoping. This phenomenon is referred to as "interference electrochromism" [23]. This phenomenon is also observed for the chiral poly(EDOT-benzene-EDOT) (PBEDOT*) [25]. The PBEDOT* shows concavo-convex structure with electro-active properties. The periodic structure based on the helical structure of the polymer produces iridescent reflection of light and diffracts light very well. The polymer film exhibits a rainbow-like appearance upon irradiation of light at certain angles (Figures 8(a)), and appears RGB monochromatic at other incident angles (Figures 8(b)). The natural colour of the polymer is dark brown in the half-doped state (Figure 8(c)). Laser transmission through the polymer film produces a circular Fourier-transformed interference with distinct separation of red, green, and blue laser light (Figure 8(d)) due to the random grating formed by the single-pitch helix-derived stripes of the polymer. The red, green, and blue wavelengths are projected in the outer, middle, and inner parts of the circular diffraction pattern, respectively. The polymer displays $m = 1$ diffraction, as shown in Figurers 8(e,g), confirming that the polymer has both transmission and reflection grating functions. Under irradiation normal to the polymer surface, the light trace draws a conic transmission and reflection pattern (Figure 8(f), graphical art). Changes in CIE colour space chromaticity upon electrochemical doping (oxidation) for the PBEDOT* film are shown in Figure 8(h,i). Ellipsometry measurements for the BEDOT* and PBTh* reveal that with increasing positive voltage, the average refractive index of the entire polymer surface decreases at short wavelengths and increases at long wavelengths. This can be regarded as a form of electrochemically driven refraction modulation.

7. Conclusions

The sandwich cell polymerisation method developed by the author allows preparation of electroactive π-conjugated polymers having both electro-activity and liquid crystal-like molecular order. The polymers thus synthesised form periodic patterns displaying periodical dielectric structure (PBTh*), fibril structure (PEDOT*), and labyrinth-like concavo-convex structure (PFFLF*, PBEDOT*). This structure organising polymerisation in LC described here allows production of structural chirality, electro-chiroptical effect, control of refraction, and interference colour tuning. For the first time, the Stille type polycondensations of achiral aromatic monomers in cholesteric LC solution were also carried out based on this idea by the author [26,27]. Furthermore, crystal-LC ordered double layer electroactive polymer has been synthesised with phase transition sequential polymerisation to show metallic electrochromism [28]. The functions derived from the combination of π-conjugated system, liquid crystallinity, chirality, and electro-activity may open a new avenue for further study of opto-electrical functional polymers. Structure-organising polymerisations by using LCs, developed in a set of the studies, can be expected for production of polymers having interesting morphology and optical properties.

Molecular imprinting (MIP) may be occurred in the polymerisation process. The chiral charge carriers in chiral polymers are defined as chiralion.

Acknowledgments

The author is deeply appreciative of the late Prof. Gursel Sonmez, great friend, for helpful discussions. The author also would like to express the appreciation to Prof. J. R. Reynolds. Mr. Jyun-ya Murakami (U. Tsukuba) helped purification of a precursor of mesogen having chiral group after the synthesis with the Mitsunobu reaction by H. Goto in 1993, as author's previous research on synthesis of ferroelectric liquid crystalline polyacetylene derivatives under supervision of Prof. Emeritus H. Shirakawa (U. Tsukuba, retired on March 2000). The author wishes to thank Prof. Akagi. Initial research on the synthesis of electrochemical polymerisation in liquid crystals and dynamically controllable chiroptical activity, and development of the polycondensation in liquid crystals were supported by the Japan Society for the Promotion of Science (JSPS), Grant-in-Aid for Scientific Research, (H. Goto, Creation of aromatic-type optically active conducting polymers by using liquid crystalline reaction field. Year: 2004–2005, No. 22550161). The present research is being supported by JSPS (2013–, No. 25410218).

References

[1] Demus D, Richter L 1978 *Texture of Liquid Crystals* (New York, Weinheim: Verlag Chemie)
[2] Bouligand Y 1972 *Tissue and Cell* **4** 189–190, 192–217
[3] Bouligand Y, Denefle J P, Lechaire J P and Maillard M *Biol. Cell* 1985 **54** 143–162
[4] Tong H, Wan P, Ma W, Zhong G, Cao L and Hu J 2008 *J. Struct. Biol.* **163** 1–9
[5] Seago A E, Brady P, Vigneron J-P and Schultz T D 2009 *J. R. Soc. Interface* **6** S165
[6] Kinoshita S and Yoshioka S eds *Structural color in biological systems–principles and applications* 2005 (Osaka: Osaka University Press)
[7] Bock H 2001 In *Chirality of Liquid Crystals*, Kitzerow H-S and Bahr C eds (New York: Springer-Verlag) p 370
[8] Goto H 1997 *Ph D thesis of U Tsukuba (in English)*
[9] Goto H 1993 *Res. Lab. Note. U. Tsukuba* **5** 5069, 0621 http://www.ims.tsukuba.ac.jp/~gotoh_lab/Note%20Goto.pdf
[10] Araya K, Mukoh A, Narahara T and Shirakawa H 1984 *Chem. Lett.* 1141
[11] Shirakawa H, Otaka T and Guanzhe P 2001 *Synth. Met.* **117** 1
[12] Piao G, Kaneko S, Higuchi I, Akagi K, Shirakawa H and Kyotani M 1999 *Synth. Met.* **101** 94
[13] Akagi K, Piao G, Shirakawa H, Kaneko S and Sakamaki K 1998 *Science* **281** 1683
[14] Goto H, 2000 *Res. Lab. Note. U. Tsukuba* **24** 2401, 2416 http://www.ims.tsukuba.ac.jp/~gotoh_lab/Note%20Goto.pdf
[15] Goto H 2004 *Res. Lab. Note. U. Tsukuba* **30** 3093 http://www.ims.tsukuba.ac.jp/~gotoh_lab/Note%20Goto.pdf
[16] Goto H and Akagi K 2002 *Book of Abstr. Int. Conf. on Sci. Tech. Synth. Met. (ICSM)*, Shanghai China (Shanghi: Reserch Center for Theoretical Physics, Fudan University) p17
[17] Goto H 2007 *Phys. Rev. Lett.* **98** 253901
[18] Goto H and Akagi K 2005 *Macromolecules* **38** 1091
[19] Huck N P M, Jaeger W F, Lange B and Fringe B L 1996 *Science* **273** 1686
[20] Goto H and Kawabata K 2011 *Polym. Chem.* **2** 1098
[21] Goto H 2010 *J. Mater. Chem.* **20** 1891
[22] Goto H 2009 *J. Mater. Chem.* **19** 4914
[23] Goto H 2009 *Adv. Funct. Mater.* **19** 1335
[24] Goto H 2010 *Mater. Chem. Phys.* **122** 69
[25] Goto H 2009 *J. Appl. Phys.* **105** 11490, Reprinted with permission, Copyright [2009], AIP Publishing LLC.
[26] Goto H 2007 *Macromolecules* **40** 1377
[27] Goto H 2003 *Res. Lab. Note. U. Tsukuba* **31** 2965, 3150 http://www.ims.tsukuba.ac.jp/~gotoh_lab/Note%20Goto.pdf
[28] Goto H 2013 *J. Polym. Sci. Part A: Polym. Chem.* **51** 3097

Spherical assemblies formed from π-conjugated alternating copolymers having fluorene and thiophene components

Yohei Yamamoto and Liang Tong

Division of Materials Science and TIMS, Faculty of Pure and Applied Sciences, University of Tsukuba, 1-1-1 Tennodai, Tsukuba, Ibaraki 305-8573, Japan.

yamamoto@ims.tsukuba.ac.jp

Abstract. Self-assembly of conducting polymers were comprehensively studied by means of slow precipitation from polymer solutions upon addition of a vapor of nonsolvents. Polymers such as polyfluorene and polythiophene hardly formed defined and discrete objects but only gave ill-defined aggregates. In contrast, some alternating copolymers having both fluorene and thiophene components in their repeating unit self-assembled into well-shaped spheres with sub-micrometer to several micrometer in the diameters. The differences in the assembling geometries derive from the crystallinity of the polymers, where the copolymers possess large steric hindrance on their backbone that reduces planarity of the polymers and inhibits anisotropic crystal growth. Changing the assembling parameters can systematically control diameter and deviation of the spheres.

1. Introduction

Construction of three-dimensional (3D) colloidal crystals have been focused much attention for obtaining novel optical properties such as optical band gap [1], nonlinear optics [2], and laser oscillation [3]. So far, various colloidal crystals consisting of beads of non-conjugated polymers such as polystyrene (PS) and poly(methyl methacrylate) (PMMA) were reported [1–3]. Meanwhile, few examples have been reported on photonic crystals consisting of π-conjugated polymers [4–7]. Examples of π-conjugated polymer photonic crystals include an inversed opal structure of poly(3-alkylthiophene) [4] and poly(*para*-phenylenevinylene) (PPV) [5,6], which are infiltrated into voids of silica or opal colloidal crystals. Another example is also from PPV spheres, prepared by an annealing of colloidal crystal of non-conjugated PPV precursors formed by so-called self-organized precipitation (SORP) method [7]. Photonic crystals consisting of π-conjugated polymers are expected to show novel optoelectronic properties such as enhanced electroluminescence, no-threshold laser oscillation, and highly-efficient photoelectric conversion as the results of charge injection, long-lived excitons, and light confinement [8,9]. Thus, developing a formation of colloidal crystals from π-conjugated polymers will take great advantages for developing new polymer-based photonic crystals.

In general, polymer colloids are prepared by either direct generation during polymerization from the corresponding monomers (direct polymerization) or secondary generation such as rapid reprecipitation or emulsification of polymers (postpolymerization). Majority of the reported π-conjugated polymer colloids were achieved by the direct polymerization techniques such as dispersion polymerization and miniemulsion polymerization [10,11]. On the other hand, not many examples are reported on spherical assemblies of π-conjugated polymers by postpolymerization [10], including self-assembly of amphiphilic π-conjugated homopolymers [12] or block copolymers [13–15]. One possible reason why π-conjugated polymers are difficult to give spherical assemblies by postpolymerization is attributed to their high rigidity and crystallinity [16,17]. π-Conjugation makes the polymer backbone planar and rigid, which lowers the flexibility of the polymers in contrast with non-conjugated polymers such as PS and PMMA. As the result of high planarity, π-conjugated polymers tend to stack on one another and crystallize anisotropically, often giving one-dimensional (1D) fibril or two-dimensional (2D) layered structures [16–19], not structurally isotropic spheres. Accordingly, one strategy for obtaining spherical assembly from π-conjugated polymers is to reduce the crystallinity, while without diminishing their electronic properties taking into consideration their optoelectronic applications.

In this proceeding, we show preparation of spherical assemblies from π-conjugated polymers by a slow precipitation from its solution upon addition of a vapor of nonsolvents. Precipitation mechanism of the vapor diffusion method is almost similar to the SORP method, in which the solvent is slowly evaporated from the polymer solution containing both 'good' and 'poor' solvents. Vapor diffusion method allows various combinations of the solvents including those having similar boiling points. We noticed that, by the vapor diffusion method, π-conjugated polymers having a single monomer component are very difficult to form well-defined spherical structures. In contrast, several alternating copolymers having two different π-conjugated molecular components tend to form quite well-shaped spheres quantitatively with its diameters ranging from a few hundreds of nanometers to several micrometers depending on the self-assembling condition.

2. Methods

Single-component polymers, **F8**, **MDMOPPV**, **rr-P3HT**, **rra-P3HT**, and alternating copolymers, **F8BT**, **PCPDTBT**, and **F8T2** (Table 1) were purchased from Aldrich Chemical Co. Ltd. Alternating copolymers, **F8TMT2** and **F8EDOT** (Table 1), were synthesized according to reported procedures [20,21]. For the preparation of self-assembled precipitates, typically, a 5-mL glass vial containing a solution of polymers (1 mg mL^{-1}, 2 mL) was placed in a 50-mL glass vial containing 5 mL of a poor solvent to allow for vapor diffusion. $CHCl_3$, CH_2Cl_2, chlorobenzene (PhCl), tetrahydrofuran (THF), and toluene were used for polymer solution, while methanol (MeOH), acetone, and hexane were used as nonsolvents. While the mixture was kept at 25 °C for 72 h, the solution turned suspension. Morphology of the precipitates was observed by scanning electron microscopy (SEM).

3. Results and Discussion

3.1. Self-assembly of π-conjugated polymers

Typical fluorescent polymers, **F8** and **MDMOPPV**, yielded pseudo-spherical objects from solvent combinations of toluene–acetone and CH_2Cl_2–MeOH, respectively, but the products were distorted in shape and heavily fused with one another. All the other combination of the solvents attempted only gave irregular aggregates or no precipitation. **rr-P3HT** gave only irregular aggregates under all solvent conditions examined. On the other hand, **rra-P3HT** formed spherical structures. However, the yielded spheres were not discrete but heavily fused with one another.

Next, we attempted self-assembly of alternating copolymers. **F8T2** having both fluorene and thiophene moieties in the repeating unit only gave ill-defined aggregates for all solvent conditions attempted. However, of interest, **F8TMT2** [20] having four methyl groups on the bithiophene unit of **F8T2** forms discrete and quite well-shaped spheres (Figure 1). The average diameter (d_{av}) and its standard deviation (σ) are 2.7 and 0.39 μm, respectively. Different solvent combinations such as $CHCl_3$ and THF for good solvents and MeOH and acetone for nonsolvents also afforded spherical assemblies, but the σ values were larger by 1.5–3 times in comparison with that obtained for a solvent combination of CH_2Cl_2/MeOH. The increase of the deviation is likely derived from the difference of the diffusion rate of the solvents. Other solvent combinations only gave ill-defined aggregate or no precipitates.

Table 1. Summary of single-component polymers and alternating copolymers, assembling condition, geometry, and size.

Polymer	Chemical Structure	M_n,[a] M_w/M_n,[a] DP[b]	Solvent/Vapor[c]	Assembling Geometry	d_{av} (σ) /μm
F8	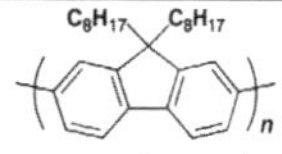	15800, 3.70, 41	Toluene /Acetone	Distorted and Fused Spheres	0.9–1.9
MDMOPPV	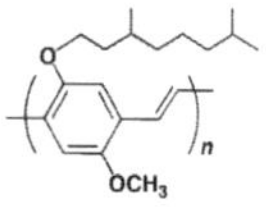	23000, –, 80	CH_2Cl_2/MeOH	Distorted and Fused Spheres	1.1–2.8
rr-P3HT (regioregular)	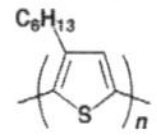	17500, –, 105	$CHCl_3$/Acetone	Irregular Aggregates	–
rra-P3HT (regiorandom)	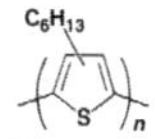	29200, 3.16, 175	$CHCl_3$/MeOH	Fused Spheres	1.8 (0.61)
F8T2	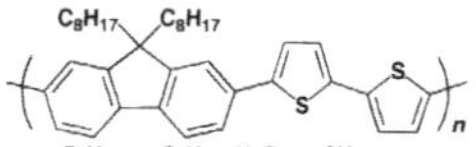	26600, 2.66, 48	$CHCl_3$/MeOH	Irregular Aggregates	–
F8TMT2	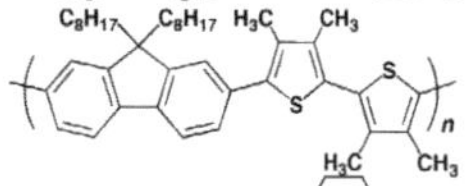	31800, 2.46, 52	CH_2Cl_2/MeOH $CHCl_3$/MeOH	Well-Shaped Spheres	2.7 (0.39) 2.4 (1.2)
F8EDOT	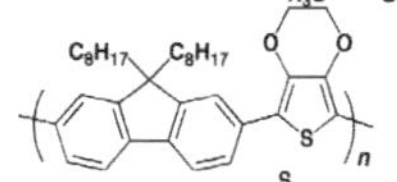	27000, 3.81, 51 13800, 2.03, 26 6300, 1.43, 12	$CHCl_3$/MeOH	Well-Shaped Spheres	0.61 (0.18) 2.2 (0.78) 4.5 (2.1)
F8BT	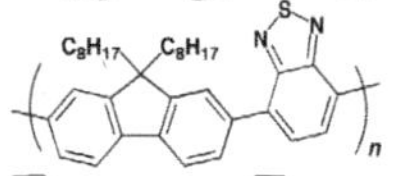	10000–20000, <3, 18–36	$CHCl_3$/Acetone	Distorted Spheres and Irregular Aggregates	4.2–7.8
PCPDTBT	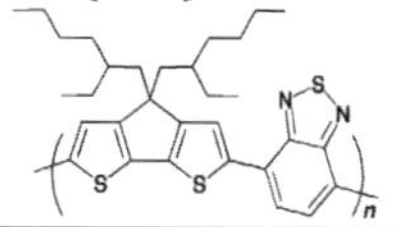	7000–20000, –, 13–37	$CHCl_3$/MeOH	Well-Shaped Spheres and Irregular Aggregates	4.5 (0.55)

[a] Estimated by gel permeation chromatography (GPC) calibrated on polystyrene standards or reported by company.
[b] Average degree of polymerization (DP), calculated from M_n and molecular weight of the repeating unit.
[c] Initial concentration and assembling temperature were 1 mg mL^{-1} and 25 °C, respectively.

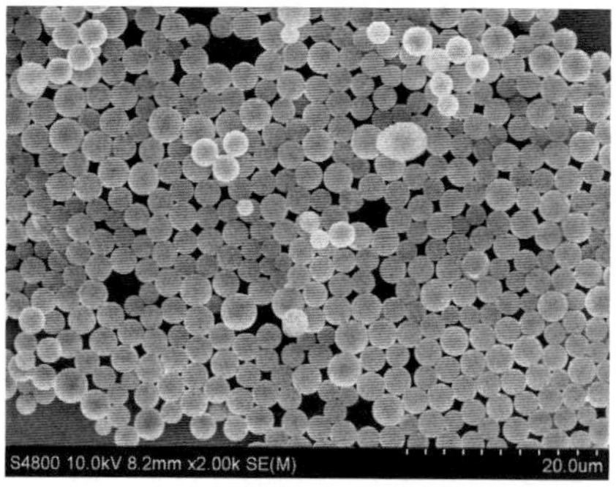

Figure 1. SEM image of air-dried $CHCl_3$/MeOH suspension of **F8TMT2**.

Similar to **F8TMT2**, an alternating copolymer, **F8EDOT** [21] having ethylenedioxythiophene (EDOT) unit instead of tetramethylbithiophene (TMT2) unit also afforded well-defined spheres with a solvent combination of $CHCl_3$/MeOH. The d_{av} and σ values were 610 and 180 nm, respectively, both of which were remarkably smaller than those formed from **F8TMT2**. Different solvent combination such as CH_2Cl_2 as a good solvent and MeOH, hexane, and acetone as poor solvents also gave spherical assemblies, but they were somewhat distorted or fused with one another. Assemblies in the other solvent combination only gave ill-defined aggregates.

We further investigated other alternating copolymers, **F8BT** and **PCPDTBT**, which are known as donor (D)–acceptor (A) polymers and often used as electron-donating layers of recent polymer photovoltaics. These polymers gave mixtures of spheres and irregular aggregates. Spheres of **F8BT** were somewhat distorted with the diameters of 4.2–7.8 μm, while those of **PCPDTBT** were well-shaped with d_{av} of 4.5 μm.

3.2. Size control of spherical assemblies

For colloidal crystals exhibiting photonic properties at the visible-light region, periodic structure of several hundreds of nanometer is required. The diameter of the spheres formed from **F8EDOT** corresponds to that length scale. Therefore, we used **F8EDOT** and investigated what factor determines the size and deviation of the spheres. **F8EDOT** having M_n of 27,000 gave spheres with d_{av} of 0.61 μm. When M_n decreased to a half (13,800), d_{av} increased to 2.2 μm, and further half of M_n (= 6,300) resulted in an increase of d_{av} to 4.5 μm. Initial concentration of the solution also affects the d_{av} values, where d_{av} of the resultant spheres decreased from 2.1 to 1.4 and 1.2 μm as the concentration of **F8EDOT** (M_n = 27,000) was increased from 0.5 to 1.0 and 2.0 mg mL^{-1}, respectively. Temperature also affects the sphere size and deviation to some extent, where aging at higher temperature resulted in the larger d_{av} values (d_{av} of 0.56, 0.67, and 0.69 μm for the spheres assembled at 0, 25, and 30 °C, respectively).

The changes of d_{av} most likely relate to the rate of precipitation. Due to the high solubility, polymers having smaller M_n precipitates with the longer aging time than those having larger M_n. Accordingly, for the small-M_n polymers, nucleation and growth take place more slowly, leading to the formation of spheres having larger diameters. Low initial concentration also results in slow nucleation and growth of spheres, giving the larger spheres in comparison with that formed from high concentration. Temperature dependency is a little complicated, because solubility is higher at higher temperature, but, at the same time, diffusion of the nonsolvent occurs more rapidly. These factors affect oppositely for the precipitation rate. In the present case, assembly at the higher temperature results in the precipitation of spheres having larger diameters, hence the former (solubility change) is the dominant factor for d_{av} of the spheres. According to these results, small-diameter spheres can be obtained by self-assembly of (i) large molecular weight polymers (ii) in high initial concentration (iii) at low temperature. Under such condition, nucleation of polymers occurs concurrently, resulting in large amount of small-size spheres.

Next, we discuss the deviation of the diameters. Spheres of **F8EDOT** with d_{av} of 0.61 μm showed 30% deviation (σ = 0.18 μm), while those with d_{av} of 4.5 μm has the larger deviation of 47% (σ = 2.1 μm). We assume that time lag of the nucleation causes the difference in the growth time of the spheres, thereby resulting in the large deviation of the diameters. In fact, **F8TMT2** in CH_2Cl_2, upon vapor diffusion of MeOH, results in the spherical precipitates with the smaller σ value than that in THF and $CHCl_3$, where evaporation of the good solvent takes place more rapidly for CH_2Cl_2 than THF and $CHCl_3$ (boiling points of CH_2Cl_2, THF, and $CHCl_3$ are 40, 66, and 61 °C, respectively). In order to reduce the time lag of the nucleation, we further demonstrated self-assemblies using a wide-mouth petri dish instead of a narrow-mouth vial. Thus, the solution/vapor interface area was increased from 1.5 to 5.7 cm^2, while the depth of the solution was reduced from 0.8 to 0.2 cm. In this case, a complete precipitation occurred within only 12 h by a vapor diffusion of MeOH into $CHCl_3$ solution of **F8EDOT**. The d_{av} and σ values of the obtained spheres were reduced by 28 and 50%, respectively, from those obtained using usual small-mouth vial (d_{av}, 0.61 → 0.44 μm; σ, 0.18 → 0.09 μm).

3.3. Mechanism for the formation of spherical assemblies

Why do spherical assemblies form upon slow diffusion of nonsolvents and why do only limited alternating copolymers assemble into such well-shaped spheres? From our results, spherical assemblies tend to form when polar nonsolvents such as MeOH and acetone are diffused into the polymer solution. These nonsolvents have low affinity with the hydrophobic polymers having linear or branched aliphatic chains. As the result, the polymers possibly aggregate while minimizing the contact area with the polar nonsolvent, thereby gradually growing into a spherical geometry as thermodynamic products. In fact, simple evaporation of the solvent from the $CHCl_3$ solution of **F8EDOT** hardly yielded spheres but only gave an amorphous film. On the other hand, when MeOH was added in advance to the $CHCl_3$ solution of **F8EDOT** ($CHCl_3$/MeOH = 10/1 v/v), and the resultant suspension was heated to dissolve the polymer and then aged at 25 °C, precipitation readily resulted within several minutes. However, the resultant precipitates were ill-shaped and fused spheres. Therefore, slow diffusion of polar nonsolvent is one of important factors to yield well-shaped spheres. In colloidal sciences, amphiphilicity with large head group and small tail group is one of driving forces for spherical assemblies [12–14, 22–24]. However, in the present case, copolymers we utilized do not possess such amphiphilicity. Instead, polarity of good- and nonsolvents plays a crucial role for the colloidal formation.

Crystallinity of polymers is another important factors. Polymers possessing high rigidity or interchain crystallinity tend to crystallize anisotropically, which disturb the assembly into isotropic spherical geometry. In fact, **rr-P3HT** having high interchain crystallinity hardly formed discrete assemblies but only gave irregular aggregates, whereas **rra-P3HT** having low interchain crystallinity due to the random regularity of the hexyl chains formed pseudo-spherical structures. Furthermore, in case of **F8TMT2** and **F8T2**, the former formed well-shaped spheres while the latter gave only ill-defined aggregates. Since four methyl groups on the bithiophene moiety in **F8TMT2** disturb a planarity of the polymer, polymer backbone most likely forms twisted configurations, leading to disordered interchain packing. On the other hand, **F8T2** possesses much planar structure, which enhances the interchain crystallinity. Copolymers **F8BT** and **PCPDTBT** also form spheres, but they are not quantitative and not well-shaped, possibly due to the rather high interchain crystallinity [25, 26].

We conducted X-ray diffraction (XRD) experiments to confirm differences of the crystallinity. However, except for **rr-P3HT**, diffraction peaks were hardly attained for all samples by convensional XRD studies, since long-range structural ordering of π-conjugated polymers are generally achieved by post-annealing of polymer films. Instead, photoemission spectra showed clear differences, giving important indication on their π-conjugation. $CHCl_3$ solutions of **F8T2** and **F8TMT2** show maximum photoluminescence peaks (λ_{PLmax}) of 499 and 468 nm, respectively, indicating that **F8T2** has the larger π-conjugation length than **F8TMT2** due to the high planarity of the polymer even in solution. The fluorescence spectra of the drop-cast film from $CHCl_3$ solution of **F8T2** exhibited significant red shift ($\Delta\lambda_{PLmax}$ = 42 nm), and irregular aggregates of **F8T2** afforded by slow precipitation showed further red-shifted fluorescence ($\Delta\lambda_{PLmax}$ = 46 nm). These results indicate that either expansion of the intrachain π-conjugation or interchain π-electronic interaction, or both of these, takes place in the solid state. In contrast, λ_{PLmax} of the cast film of the spheres of **F8TMT2** (475 nm, $\Delta\lambda_{PLmax}$ = 7 nm), as well as that formed from $CHCl_3$ solution of **F8TMT2** (474 nm, $\Delta\lambda_{PLmax}$ = 6 nm), were not so much shifted from λ_{PLmax} of the solution of **F8TMT2**. These results clearly indicate that neither expansion of intrachain π-conjugation nor interchain π-electronic interaction take place in the solid films of **F8TMT2**, irrespective of whether the aggregation takes place kinetically (solution-cast) or thermodynamically (spheres). Concerning **F8EDOT**, a certain red shift of λ_{PLmax} was observed by the formation of spheres ($\Delta\lambda_{PLmax}$ = 37 nm), which was rather large in comparison with that of **F8TMT2** ($\Delta\lambda_{PLmax}$ = 6 nm) but not as large as that of **F8T2** ($\Delta\lambda_{PLmax}$ = 46 nm). We assume that the degree of steric hindrance of the ethylenedioxy moieties in **F8EDOT** is smaller than that of the methyl groups in **F8TMT2**, which results in the smaller twisting of the π-conjugation in **F8EDOT** than **F8TMT2**. These results also support the fact that **F8TMT2** easily forms well-defined spheres under various solvent combinations whereas **F8EDOT** forms spheres only in a limited solvent combination.

4. Conclusion

We studied self-assembly of π-conjugated polymers upon slow precipitation from their solution. While most representative π-conjugated polymers hardly yielded spherical assemblies, some alternating copolymers assembled into well-shaped spherical structures quantitatively. The main factor for spherical assembly is attributed to the low crystallinity of the polymers, where copolymers with steric hindrance in their backbone and thereby having low crystallinity tend to form spheres. Optimizing the self-assembling conditions can systematically control size and deviation of the spheres. Photocarrier lifetimes in the spheres were markedly enhanced in comparison with that of the samples without forming spheres. Yet more attempts are needed to create colloidal crystals such as precise control of the size and deviation, this research provides important knowledge for photonic applications using π-conjugated functional polymers.

Acknowledgments

The authors acknowledge Prof. Takaki Kanbara and Prof. Junpei Kuwabara in University of Tsukuba for synthesis of alternating copolymers, **F8EDOT** and **F8TMT2**, and Prof. Yoshikazu Suzuki in University of Tsukuba for XRD measurements.

References

[1] Kanai T, Sawada T, Toyotama, A and Kitamura K 2005 *Adv. Funct. Mater.* **15** 25
[2] Molinos-Gómes A, Maymó M, Vidal X, Velasco D, Martorell J and López-Calahorra, F 2007 *Adv. Mater.* **19** 3814
[3] Furumi S, Kanai T and Sawada T 2011 *Adv. Mater.* **23** 3815
[4] Sato S, Kajii H, Kawagishi Y, Fujii A, Ozaki M and Yoshino K 1999 *Jpn. J. Appl. Phys.* **38** L1475
[5] Polson R. C, Chipouline A and Vardeny Z. V. 2001 *Adv. Mater.* **13** 760
[6] Deutsch M, Vlasov Y. A, Norris D J 2000 *Adv. Mater.* **12** 1176
[7] Yabu H, Tajima A, Higuchi T and Shimomura M 2008 *Chem. Commun.* 4588
[8] John S and Quang T 1997 *Phys. Rev. Lett.* **78** 1888
[9] Yoshino K, Tatsuhara S, Kawagishi Y, Ozaki M and Zakhidov A A 1999 *Appl. Phys. Lett.* **74** 2590
[10] Pecher J and Mecking S 2010 *Chem. Rev.* **110** 6260
[11] Kuehne A J C, Gather M C and Sprakel J 2012 *Nat. Commun.* **3** 1088
[12] Alam M M, Zhu Y and Jenekhe S A 2003 *Langmuir* **19** 8625
[13] Jenekhe S A and Chen X L 1998 *Science* **279** 1903
[14] Jenekhe S A and Chen X L 2000 *J. Phys. Chem. B* **104** 6332
[15] Chen X L and Jenekhe S A 2000 *Macromolecules* **33** 4610
[16] Beaujuge P M and Fréchet J M J 2011 *J. Am. Chem. Soc.* **133** 20009
[17] Kim F S, Ren G and Jenekhe S A 2011 *Chem. Mater.* **23** 682
[18] Sirringhaus H, Brown P J, Friend R H, Neilsen M M, Bechgaard K, Langeveld-Voss B M W, Spiering A J H, Janssen R A J, Meijer E W, Herwig P and de Leeuw D M 1999 *Nature* **401** 685
[19] Yang X, Loos J, Veenstra S C, Verhees W J H, Wienk M W, Kroon J M, Michels M A J and Janssen R A 2005 J. *Nano Lett.* **5**, 579
[20] Fujinami Y, Kuwabara J, Lu W, Hayashi H, Kanbara T 2012 *ACS Macro Lett.* **1** 67
[21] Yamazaki K, Kuwabara J, Kanbara T 2013 *Macromol. Rapid Commun.* **34** 69
[22] Fujita N, Yamashita T, Asai M and Shinkai S 2005 *Angew. Chem. Int. Ed.* **44** 1257
[23] Koizumi Y, Seki S, Tsukuda S, Sakamoto S and Tagawa S 2006 *J. Am. Chem. Soc.* **128**, 9036
[24] Nakanishi T, Michinobu T, Yoshida K, Shirahata N, Ariga K, Möhwald H and Kurth D G 2008 *Adv. Mater.* **20** 443
[25] Zhang M, Tsao H N, Pisula W, Yang C, Mishra A K and Müllen K 2007 *J. Am. Chem. Soc.* **129**

3472
[26] Tsao H N, Cho D, Andreasen J W, Rouhanipour A, Breiby D W, Pisula W and Müllen K 2009 *Adv. Mater.* **21** 209

IOP Conf. Series: Materials Science and Engineering **54** (2014) 012015 doi:10.1088/1757-899X/54/1/012015

Triphenylamine-based amorphous polymers for bulk-heterojunction photovoltaic cells

Takeshi Yasuda[1]
Organic Thin-Film Solar Cells Group, Photovoltaic Materials Unit, National Institute for Materials Science (NIMS), 1-2-1 Sengen, Tsukuba, Ibaraki 305-0047, Japan

E-mail: YASUDA.Takeshi@nims.go.jp

Abstract. In this paper, the recent research progress on triphenylamine (TPA)-based donor-acceptor (D-A) amorphous polymers including our developed polymers is reviewed. TPA has three-dimensional branched structures and can provide D-A polymers containing D and A units in the main chain or side chain. The use of TPA-based amorphous polymers in the fabrication of organic photovoltaics (OPVs) offers great advantages over the use of a polycrystalline film in terms of high reproducibility of the OPV performance. The amorphous polymer design using TPA, therefore, indicates a promising direction for the development of new donor materials in OPVs.

1. Introduction

Organic photovoltaics (OPVs) based on solution-processed bulk-heterojunctions (BHJs), where π-conjugated polymer (donor) and fullerene derivative (acceptor) phases are separated during solution coating to form nanoscale donor/acceptor interfaces, have garnered considerable research interest because of their significant advantages such as their low cost, low weight, and mechanical flexibility [1-3]. Interdisciplinary research into BHJ OPV has resulted in a rapid enhancement in the power conversion efficiency (PCE) values by more than 8% under standard illumination conditions (AM 1.5) [4,5]. The development of novel conjugated polymers for BHJ OPV cells has contributed to the significant improvements in PCEs. In recent years, efforts have mainly been focused on varying the molecular structure of π-conjugated polymer backbones. For example, low-bandgap polymers are obtained with an alternating electronic donor-acceptor (D-A) structure along the backbone [6-8]. The presence of D and A units on one chain creates a push-pull charge-transfer (CT) effect that can be excited by lower-energy photons, leading to a red-shifted onset of absorption. This approach has been used to develop efficient, low-bandgap photovoltaic materials. Due to successful applications in organic light-emitting diodes (OLEDs) and organic thin-film transistors (OTFTs), fluorene, carbazole, thiophene and their analogue heterocycles have become the first choice as donors in BHJ OPVs.

Recent years have seen the emergence of a new approach based on the use of triphenylamine (TPA) units in the design of amorphous materials that are used as donors in BHJ solar cells. These TPA-based amorphous polymers present a number of specific advantages, due to their excellent thermal and electrochemical stability, electron donating ability, hole transport properties, and high absorption coefficients. They are now known to be important hole transport layer (HTL) materials in the commercial production of OLEDs [9]. Ultrathin films composed of a pinhole-free homogeneous layer

[1] To whom any correspondence should be addressed.

with a thickness of less than 100 nm are suitable for HTL for OLEDs as well as for OPVs. Moreover, the incorporation of TPA units into π-conjugated polymers is known to be quite effective for the control of HOMO-LUMO levels and absorption spectra, as well as of their hole transport capability [10]. Thus, the use of TPA units in the design of polymer skeletons for not only OLEDs but also OPVs is an attractive option.

In this paper, the recent research progress on TPA-based D-A amorphous polymers including our developed polymers is reviewed. This review is organized as follows: an overview of TPA-based polymers containing D and A units in the main chain or side chain in the section 2. In the subsequent section 3, an advantage of using TPA-based amorphous polymers in the fabrication of OPVs is described.

2. Designs and OPV properties of triphenylamine-based D-A polymers

As mentioned in the introduction, to obtain low-bandgap polymers, one feasible approach is to design alternating D-A copolymers. TPA has three-dimensional branched structures and can provide D-A polymers containing D and A units in the main chain (Figure 1) or side chain (Figure 2). So, I introduce TPA-based D-A polymers divided into the two structures.

2.1. Main-chain D-A polymers

A variety of electro-accepting units such as benzothiadiazole (BTD), dithienylbenzothiadiazole (TBT), and diketopyrrolopyrrole (DPP), have been employed as the counterpart in designing D-A polymers. The author's group reported a comparative study of the properties of BHJ OPVs using a BTD-TPA small molecule [11] and its polymerized molecule (**P1**) [12]. OPVs using BTD-TPA or **P1**:$PC_{60}BM$ at a 1:2 mixing weight ratio were fabricated. The PCE of the **P1**-based OPV was twice that of the BTD-TPA-based OPV. The field-effect hole mobility of **P1** (3.1×10^{-4} $cm^2V^{-1}s^{-1}$) is two orders of magnitude higher than that of BTD-TPA (5.4×10^{-6} $cm^2V^{-1}s^{-1}$) and the absorption peak of **P1** (494 nm) is at a longer wavelength than that of BTD-TPA (477 nm). Accordingly, the improved hole mobility and enhanced absorption of AM 1.5 solar-simulated light led to a high short-circuit current (J_{sc}) and PCE in OPVs based on **P1**. By optimizing OPVs with **P1**:$PC_{70}BM$ (1:4), the device performance exhibited a J_{sc} value of 7.45 $mAcm^{-2}$, an open-circuit voltage (V_{oc}) of 0.92 V, a PCE of 2.65%, and incident photon to current conversion efficiencies (IPCEs) of around 50% at wavelengths ranging from 360 to 560 nm.

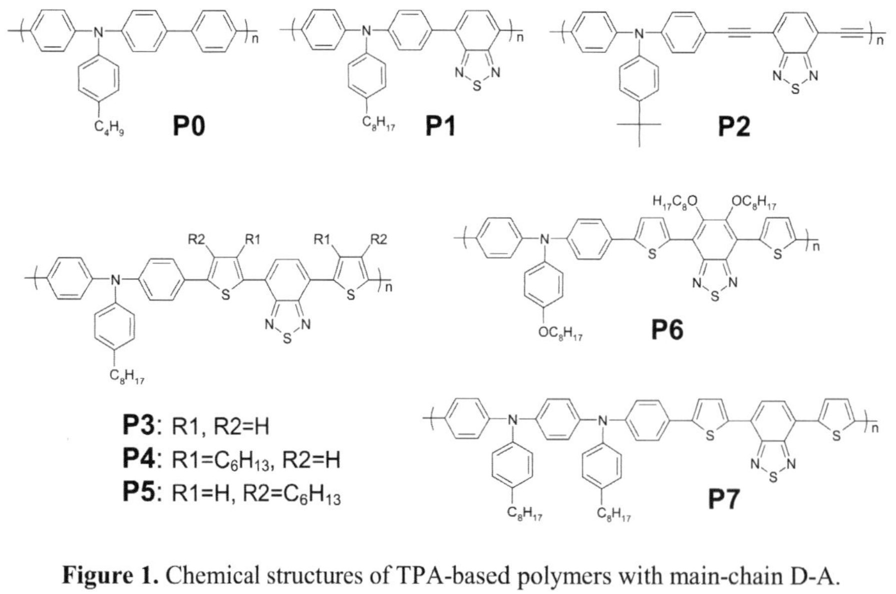

Figure 1. Chemical structures of TPA-based polymers with main-chain D-A.

Blankenburg et al. reported the synthesis of **P2** in which BTD and TPA are connected together with rigid acetylenic linkages and the investigation of their optical, electrochemical, and photovoltaic properties [13]. The copolymer was synthesized by the standard Sonogashira-reaction of selected dihalogen BTD with diethynyl TPAs. The controlled polymers in which thiophene or 3-alkyl-thiophene or alkoxy phenylene was connected with diethynyl TPAs show a prominent absorption at 408-411 nm. Compared with these polymers, the integration of BTD (**P2**) led to the expected red shift of approximately 100 nm. PCE of 1.85% could be reached without any optimization of concentration of the solutions, the ratio of the polymer to PCBM, the spin coating process itself or annealing.

The steric hindrance between the acceptor BTD unit and the donor benzene-based aromatic units has previously been reported [14], and the conjugated backbone in **P1** is also twisted at the BTD-TPA linkage, leading to the possibility of reduced effective conjugation. So, some BTD-TPA-based polymers with a more red-shifted onset of absorption by introducing flanking thienyl groups between BTD and TPA to reduce the amount of steric hindrance between them have been developed [15]. The TBT-TPA polymers, **P3**, **P4** and **P5** differ only by the presence and position of the side chains on the thienyl groups, as shown in Figure 1. **P3** has no side chains on the flanking thienyl groups, **P4** has hexyl side chains on the three-positions of the thienyl groups (pointing toward the BTD unit), and **P5** has hexyl side chains on four-positions of the thienyl group (pointing outward from the BTD unit).

The obtained **P4** and **P5** are soluble in common organic solvents including chloroform ($CHCl_3$) and *o*-dichlorobenzene (*o*-DCB). However, **P3** was produced as an insoluble material during the polymerization, so the additional thiophene rings could result in a low weight average molecular weight (M_w= 3400) for **P3** because of its low solubility. The additional thiophene rings with hexyl side-chains between the BTD-TPA backbones, on the other hand, were attached to engender good solubility and led to an increase in M_w to 21,900. By the addition of alkoxy groups to BTD and two phenyl amines, other TBT-based polymers, **P6** and **P7,** have good solubility in common organic solvents [16,17].

The UV–vis absorption maxima in the CT absorption spectra for the **P4** and **P5** films appeared at 501 and 538 nm, respectively. The UV-vis absorption maxima for the **P1** film appeared at 494 nm. Therefore, BTD-TPA based polymers with a more red-shifted onset of absorption by introducing the flanking thienyl groups between the BTD and TPA have been developed. A comparison of spectra for the **P4** and **P5** films reveals that the former has a slightly blue-shifted and weakened CT absorption. The spectral position and relative intensity of these peaks varied strongly with the change in the side-

Table 1. Photovoltaic parameters for BHJ OPV illuminated at 100 mWcm^{-2} of AM 1.5.

Polymer	HOMO (-eV)	Absorption in film (nm)	BHJ layer	J_{sc} (mAcm^{-2})	V_{oc} (V)	FF	PCE (%)	Ref.
P0	5.33[a]	379						10
P1	5.47[a]	494	**P1**:$PC_{70}BM$ (1:4)	7.45	0.92	0.39	2.65	12
P2	5.42[b]	503[c]	**P2**:$PC_{60}BM$ (1:4)	5.22	0.89	0.40	1.85	13
P4	5.36[a]	501	**P4**:$PC_{70}BM$ (1:4)	6.50	0.83	0.39	2.12	15
P5	5.26[a]	538	**P5**:$PC_{70}BM$ (1:4)	6.48	0.86	0.38	2.10	15
P6	5.29[b]	558	**P6**:$PC_{70}BM$ (1:3)	6.9	0.77	0.43	2.3	16
P7	4.8[b]	577	**P7**:$PC_{70}BM$ (1:3)	3.8	0.62	0.48	1.0	17
P8	5.30[b]	548	**P8**:**PC_{70}**BM (1:4)	7.00	0.84	0.42	2.44	18
P9	5.32[b]	554	**P9**:$PC_{70}BM$ (1:4)	7.4	0.93	0.44	3.1	19
P10	5.38[b]	518	**P10**:$PC_{70}BM$ (1:3)	9.40	0.86	0.39	3.09	20
P11	5.22[b]	653	**P11**:$PC_{70}BM$ (1:3)	10.77	0.80	0.35	2.95	20

[a] The values were estimated by photoelectron yield spectroscopy.
[b] The values were obtained from cyclic voltammograms.
[c] Absorption in a solution state.

chain position on the thiophene unit. When the hexyl chains on thiophene rings face toward the BTD, severe steric hindrance is introduced between the flanking thienyl groups and the central BTD unit. In such a case, the conjugated backbone is thereby twisted at the D-A linkage. This leads to a blue-shifted and weakened CT absorption. IPCEs in the region from 550 to 650 nm for the OPV cells prepared using **P5** were higher than those for the OPV cells prepared using **P4** because the absorption spectrum for the **P5** has a slightly red-shifted absorption. The photovoltaic properties of **P1-P7** are summarized in Table 1.

2.2. Side-chain D-A polymers

Recent studies have developed a new family of conjugated polymers with conjugated side chains (Figure 2). This type of polymers features high hole mobility thanks to the overlapping of the side chain interactions with the conjugated main chains, and broad absorptions deriving from both the main chains and the conjugated side chains.

As successful examples, Zhang et al. reported copolymers of carbazole and TPA with thienylenevinylene conjugated side chain containing a dicyano acceptor end group (**P8**) [18]. BHJ OPV based on **P8**:$PC_{70}BM$ (1:4) demonstrated OPV performance with J_{sc} of 7.0 $mAcm^{-2}$, FF of 0.42, V_{oc} of 0.84 V, corresponding to a PCE of 2.44%. Duan et al. also reported copolymers of indenofluorene and TPA with the same side chain (**P9**) and the OPV performance with J_{sc} of 7.4 $mAcm^{-2}$, FF of 0.44, V_{oc} of 0.93 V, corresponding to a PCE of 3.1% [19].

Nie et al. developed two side-chain D-A conjugated copolymers (**P10** and **P11**), in which an alternating binary donor units of benzodithiophene and TPA conjugatedly formed the main chain, and TBT or DPP acceptor units were pended onto the TPA units to build the side chain [20]. The choice of alternating binary donor units and pending acceptor unit was expected to broaden the absorption area

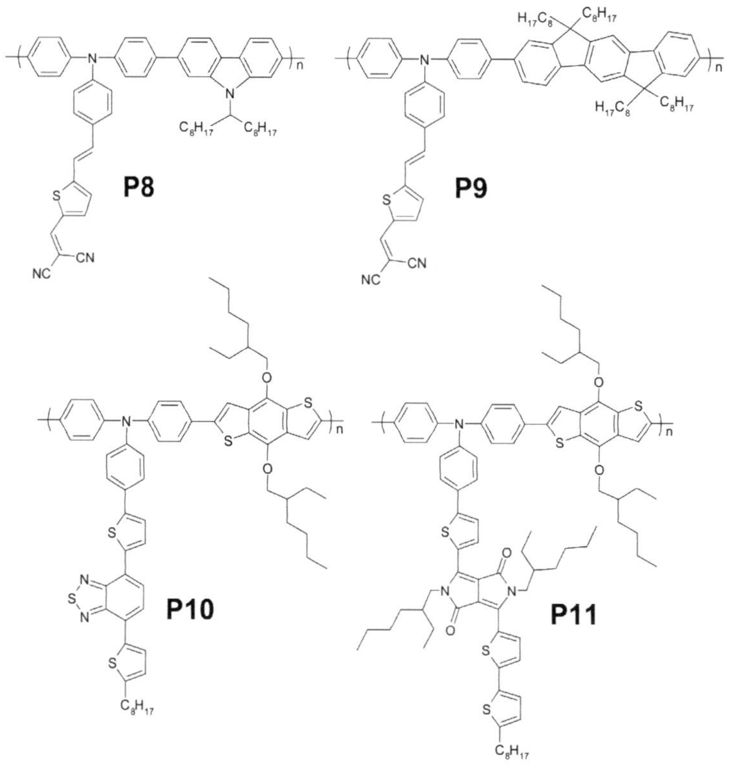

Figure 2. Chemical structures of TPA-based polymers with side-chain D-A.

for their resulting copolymers, which was critical to improve device performance. The absorption peaks at 518 nm for **P10** and at 653 nm for **P11** were observed in their films, respectively. It indicated that attaching the DPP unit instead of TBT unit into the side chain can effectively tune the absorption properties. Compared to the **P10**-based OPVs, the **P11**-based OPVs presented an increasing J_{sc} of 10.77 $mAcm^{-2}$. The photovoltaic properties of **P8-P11** are summarized in Table 1.

3. Advantage of using TPA-based amorphous polymers in the fabrication of OPVs

As mentioned in the introduction, remarkable improvements in the performance of BHJ OPVs have been made in recent years. In contrast, it is insufficiently known and inadequately documented that the performance of BHJ OPVs based on polycrystalline polymers such as regioregular poly(3-hexylthiophene) (P3HT) is largely dependent on the fabrication conditions such as the solvents used for dissolving the polymer and PCBM, the film-fabrication methods, and the annealing temperatures used for the BHJ films. Large fluctuations in the PCEs from sample to sample are observed even in BHJ OPVs using the same active layers [21]. To ensure proper industrial application, the PCE deviation of fabricated BHJ OPVs should be as small as possible.

The use of an amorphous film in the fabrication of OPVs offers great advantages over the use of a polycrystalline film in terms of high reproducibility of the OPV performance; TPA-based OPV performances are independent of the casting solvent and the thermal annealing temperature. The OPV based on **P1**:$PC_{70}BM$ (1:4) that was fabricated using $CHCl_3$ (boiling point of 61 °C) and annealed at 60 °C for 10 min exhibited a PCE of 2.81%. On the other hand, the OPV fabricated using *o*-DCB (boiling point of 181 °C) and annealed at 110 °C for 10 min exhibited a PCE of 2.65%. Almost the same PCEs were obtained in both OPVs [22].

The performance of BHJ-OPV using **P5** was also independent of the casting solvent and the thermal annealing temperature. The OPV based on **P5**:$PC_{70}BM$ (1:4) that was fabricated using the $CHCl_3$ solution and annealed at 60 °C for 10 min exhibited a PCE of 2.09%. The OPV fabricated using the *o*-DCB solution and annealed at 110 °C for 10 min, conversely, exhibited a PCE of 2.10% [15]. Moreover, Wang et al. reported the OPVs of **P6**:$PC_{70}BM$ (1:3) fabricated from trichlorobenzene (TCB, boiling point of 208 °C) or chlorobenzene (CB, boiling point of 132 °C) solution both showed PCE of 2.1% [16]. The reproducible photovoltaic properties of these polymers are summarized in Table 2.

Table 2. Processes of fabricating BHJ layers and the photovoltaic parameters under illumination with 100 $mWcm^{-2}$ of AM 1.5.

Polymer	Solvent	Annealing temperature (°C)	BHJ layer	J_{sc} ($mAcm^{-2}$)	V_{oc} (V)	FF	PCE (%)	Ref.
P1	*o*-DCB	110	**P1**:$PC_{70}BM$ (1:4)	7.45	0.92	0.39	2.65	22
P1	$CHCl_3$	60	**P1**:$PC_{70}BM$ (1:4)	8.48	0.93	0.39	2.81	22
P5	*o*-DCB	110	**P5**:$PC_{70}BM$ (1:4)	6.48	0.86	0.38	2.10	15
P5	$CHCl_3$	60	**P5**:$PC_{70}BM$ (1:4)	6.49	0.85	0.38	2.09	15
P6	TCB	-	**P6**:$PC_{70}BM$ (1:3)	5.4	0.77	0.50	2.1	16
P6	CB	-	**P6**:$PC_{70}BM$ (1:3)	5.7	0.76	0.46	2.1	16

4. Concluding remarks

TPA has three-dimensional branched structures and can provide D-A polymers containing D and A units in the main chain or side chain. The use of amorphous TPA-based polymers in the fabrication of OPVs offers great advantages over the use of a polycrystalline film in terms of high reproducibility of the OPV performance. It is striking that the performance of the solar cells are not extremely sensitive to the solvents, which can simplify the preparation of devices and it is unnecessary to select one kind of the solvent strictly. Such insensitivity might be due to three-dimensional branched structures of TPA-based polymers, since the structure can suppress the aggregation of polymer chains. The amorphous TPA-based polymer design, therefore, indicates a promising direction for the development of new donor materials in OPVs.

References

[1] Li G, Zhu R and Yang Y 2012 *Nature Photon.* **6** 153
[2] Nielsen T. D, Cruickshank C, Foged S, Thorsen J and Krebs F. C 2010 *Sol. Energy Mater. Sol. Cells* **94** 1553
[3] Krebs F. C 2009 *Sol. Energy Mater. Sol. Cells* **93** 394
[4] Letian D, Chen C. C, Yoshimura K, Ohya K, Chang W. H, Gao J, Liu Y, Richard E and Yang Y 2013 *Macromolecules* **46** 3384
[5] He Z, Zhong C, Su S, Xu M, Wu H and Cao Y 2012 *Nature Photon.* **6** 591
[6] Cai S, Chen L, Zha D and Chen Y 2013 *J. Polym. Sci. Part A: Polym. Chem.* **51** 624
[7] Cui C, Fan H, Guo X, Zhang M, He Y, Zhan X and Li Y 2012 *Polym. Chem.* **3** 99
[8] Nakanishi T, Shirai Y, Yasuda T and Han L 2012 *J. Polym. Sci. Part A: Polym. Chem.* **50** 4829
[9] Shirota Y and Kageyama H 2007 *Chem. Rev.* **107** 953
[10] Yasuda T, Suzuki T, Takahashi M and Tsutsui T 2009 *Chem. Lett.* **38** 1040
[11] Yasuda T, Ishi-i T and Han L 2010 *J. Photopolym. Sci. Tech.* **23** 307
[12] Yasuda T, Shinohara Y, Matsuda T, Han L and Ishi-i T 2012 *J. Mater. Chem.* **22** 2539
[13] Blankenburg L, Hager M D, Sell S, Sensfuss S and Klemm E 2009 *J. Appl. Polym. Sci.* **111** 1850
[14] Zhou H, Yang L, Xiao S, Liu S and You W 2010 *Macromolecules* **43** 811
[15] Yasuda T, Shinohara Y, Matsuda T, Han L and Ishi-i T 2013 *J. Polym. Sci. Part A: Polym. Chem.* **51** 2536
[16] Wang M, Li C, Lv A, Wang Z, Bo Z and Zhang F 2012 *Polymer* **53** 324
[17] Tu G, Massip S, Oberhumer P. M, He X, Friend R. H, Greenham N. C and Huck W. T. S 2010 *J. Mater. Chem.* **20** 9231
[18] Zhang Z. G, Liu Y. L, Yang Y, Hou K, Peng B, Zhao G, Zhang M, Guo X, Kang E. T and Li Y 2010 *Macromolecules* **43** 9376
[19] Duan C, Cai W, Zhong C, Li Y, Wang X, Huang F and Cao Y 2011 *J. Polym. Sci. Part A: Polym. Chem.* **49** 4406
[20] Nie K, Tan H, Deng X, Wang Y, Chen Q, Huang Y, Liu Y, Yang C, Huang Z, Zhu M and Zhu W 2013 *J. Polym. Sci. Part A: Polym. Chem.* **51** 4103
[21] Dang M. T, Hirsch L and Wantz G 2011 *Adv. Mater.* **23** 3597
[22] Yasuda T, Shinohara Y, Ishi-i T and Han L 2012 *Org. Electron.* **13** 1802

IOP Conf. Series: Materials Science and Engineering **54** (2014) 012016 doi:10.1088/1757-899X/54/1/012016

Crystalline conjugated polymers for organic electronics

Itaru Osaka and Kazuo Takimiya

RIKEN Center for Emergent Matter Science, Wako, Saitama 351-0198, Japan

E-mail: itaru.osaka@riken.jp

Abstract. We reported the characterization of a series of semiconducting polymers based on naphthodithiophenes (NDTs) and naphthobisthiadiazole (NTz). Four different isomeric NDTs are selectively synthesized using the similar synthetic protocol. The isomeric NDTs provide quite distinct electronic structures and crystallinity of the corresponding polymers and thus the device performances. Polymers with one of the NDT isomers, NDT3, showed high hole mobilities in excess of 0.5 cm^2/Vs. On the other hand, an NTz-based polymer exhibited high hole mobility of >0.5 cm^2/Vs in OFETs and power conversion efficiencies of >7% in OPVs. These high performances of the present polymers are found to originate in their highly crystalline structures in the thin films.

1. Introduction

With their great electrical and optical properties and ease of fabrication process using their solutions, π-conjugated polymers, namely, semiconducting polymers have been attracting considerable attentions for applications to various organic electronic devices such as field-effect transistors (OFETs) and photovoltaics (OPVs) [1-2]. A key to develop high-performance semiconducting polymers is to achieve strong π–π stacking structures that determines the charge transport property in the active layer films [3-6]. Therefore, the incorporation of π-cores into the backbones and/or the use of donor–acceptor backbones that can promote strong intermolecular interaction are important design strategies [1]. In parallel, understanding the structure–property relationships, i.e., correlation between polymer structures and their optical, physical, and device properties, is also crucial for the development of polymers.

In this paper, we reported on a series of semiconducting polymers based on naphthodithiophenes (NDTs) [7-11] and naphthobisthiadiazoles (NTz) [12], both four-ring-fused heteroaromatics. These polymers afford higher crystalline structures in the thin films, which are beneficial to OFETs and OPVs.

2. Naphthodithiophene-based polymers

Figure 1 shows the chemical structure of the NDT-based polymers. All polymers were soluble in hot chlorobenzene, *o*-dichlorobenzene, and 1,2,4-trichlorobenzene, where the solubility was PNDT2BT > PNDT1BT > PNDT3BT > PNDT4BT from the highest, and the polymers with $C_{20}H_{41}$ (C20) side chains showed better solubility than those with $C_{16}H_{33}$ (C16) side chains. Molecular weights (M_n) of the polymers evaluated by high-temperature (140 °C) GPC calibrated with polystyrene standard were 21000–52000 g/mol.

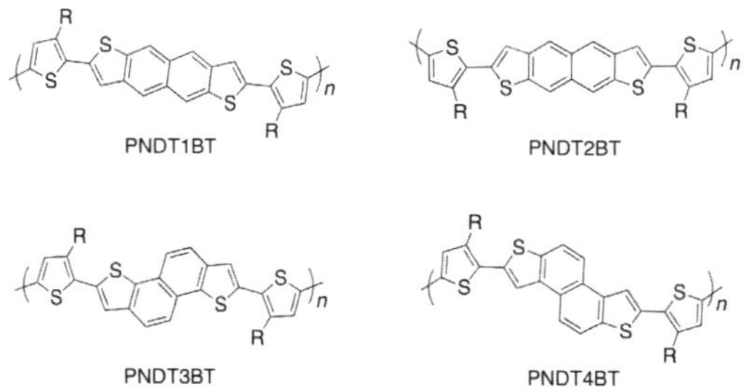

Figure 1. Chemical structures of NDT-based semiconducting polymers.

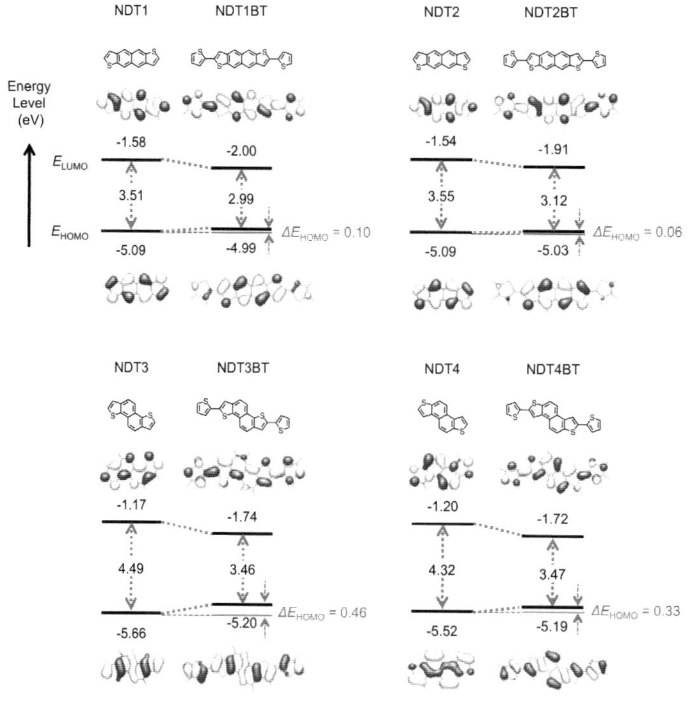

Figure 2. Calculated HOMOs and LUMOs of NDTs and the model compounds (NDT*m*BTs).

In the UV-vis absorption spectra, room temperature solutions of PNDT1BT-16, PNDT3BT-16, and PNDT4BT-16 gave two peaks at 532 and 571 nm, 506 and 540 nm, and 492 and 529 nm, respectively, which are mostly identical with those in the as-spun thin film. PNDT2BT-16, on the other hand, gave a single peak with λ_{max} of 487 nm in the solution, which slightly broadens toward the longer wavelength region in the as-spun thin film. Photoelectron spectroscopy in air was carried out to evaluate the ionization potential (IP) of the polymers in the thin film. The IPs were 5.1, 5.2, 5.0, and 5.2 eV for PNDT1BT–PNDT4BT, respectively. Interestingly, the IP of PNDT1BT was larger than that of PNDT3BT, and similar to PNDT4BT. Interestingly IP of PNDT1BT was larger than that of PNDT3BT despite the fact that PNDT1BT had a smaller bandgap than PNDT3BT. A plausible explanation can be described by taking into account the calculated HOMOs of corresponding model compounds (Figure 2). In PNDT1BT, the HOMOs are apparently *localized* within the NDT cores,

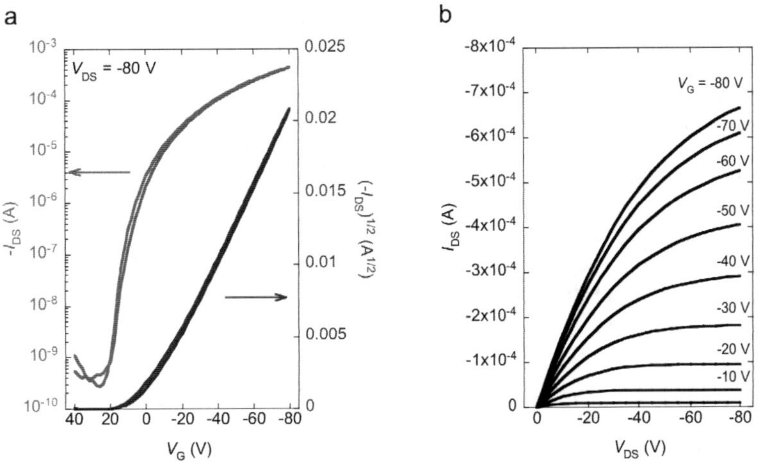

Figure 3. (a) Transfer and (b) output curves of an OFET device with PNDT3BT-20 after annealing.

Table 1. Mobilities and on/off ratios

Polymer	Side chain	μ (cm^2 V^{-1} s^{-1})/ maximum (average)		Ion/Ioff
		HMDS	FDTS	
PNDT1BT	C16	0.029 (0.025)	0.024 (0.020)	~106
	C20	0.037 (0.027)	0.055 (0.032)	~104
PNDT2BT	C16	0.0059 (0.0039)	0.0033 (0.0019)	~105
	C20	0.0066 (0.0037)	0.026 (0.0016)	~103
PNDT3BT	C16	0.54 (0.20)	0.26 (0.12)	~107
	C20	0.32 (0.18)	0.77 (0.38)	~107
PNDT4BT	C16	0.086 (0.043)	0.030 (0.022)	~107
	C20	0.15 (0.093)	0.19 (0.14)	~107

whereas in PNDT3BT the HOMOs are *delocalized* over the whole molecule structures (including attached thiophene rings). Therefore, the elevation of the E_{HOMO} ($\varDelta E_{HOMO}$) from the parent NDTs to NDTmBTs is more significant in the angular NDT system as compared to the linear NDT system; for example, in the NDT1 system $\varDelta E_{HOMO}$ (= |E_{HOMO} (NDTm) – E_{HOMO} (NDTmBT) |) is ca. 0.10 (=|-5.09 – (-4.99)|) eV, whereas in the NDT3 system $\varDelta E_{HOMO}$ is ca. 0.46 (=|-5.66 – (-5.20)|) eV. Assuming that this estimation can be adapted in the actual polymer system, the E_{HOMO} of the linear and angular NDT systems would get closer as the molecular chains extend to end up with the similar or possibly reversed E_{HOMO}, which well agrees with the results observed for the present polymer system.

OFET characteristics of the polymers were evaluated by top contact, bottom gate devices fabricated by using polymer thin films spin-coated from *o*-dichlorobenzene solutions onto hexamethyldisilazane (HMDS)- and *1H,1H,2H,2H*-perfluorodecyltriethoxysilane (FDTS)-modified Si/SiO_2 surfaces, respectively, which were subsequently annealed at 150 °C. Mobilities were calculated from the saturation regime and are summarized in Table 1. PNDT3BT-20 on the FDTS-modified devices exhibited the highest mobility of 0.77 $cm^2V^{-1}s^{-1}$ (average 0.38 $cm^2V^{-1}s^{-1}$) with high on/off ratios of ~10^7 (Figure 3). This value is among the highest observed so far for semiconducting polymers. The mobilities obtained for these polymers are PNDT3BTs > PNDT4BTs > PNDT1BTs > PNDT2BTs, e.g., the average mobility of the C20 polymers on the HMDS-modified devices are 0.18, 0.093, 0.027,

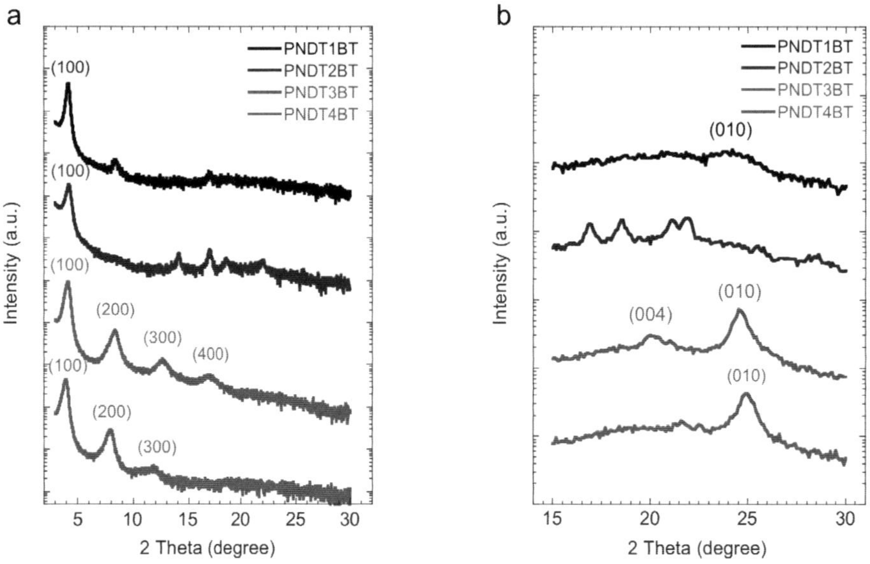

Figure 4. Out-of-plane (a) and in-plane (b) GIXRD patterns of PNDT*m*BT-16.

0.0037 $cm^2V^{-1}s^{-1}$, respectively. This trend is quite different from our initial expectation; we expected that PNDT1BT would give the best performance, since NDT1 showed the highest mobility among the four NDT isomers in the small molecular systems. To understand the difference of the mobility in these polymers and their unexpected trend, we carefully studied the ordering structures using grazing incidence X-ray diffraction (GIXRD) measurements.

The out-of-plane GIXRD patterns gave sharp peaks assignable to (*h*00), where (100) peaks appeared at around $2\theta = 4.2°$, *d*-spacings (d_l) of 20–22 Å, which correspond to the lamellar structure along with side chain interdigitation (Figure 4a). Since these (*h*00) peaks were not seen in the in-plane patterns, the polymers are dominantly edge-on oriented respective to the substrate. It was found that, interestingly, each polymer gave quite distinct diffraction features in the out-of-plane direction. PNDT3BT-16 provids (*h*00) peaks up to the fourth order, the third order for PNDT4BT-16, the second order for PNDT1BT-16, and only the first order for PNDT2BT-16, indicating that PNDT3BT-16 forms the most highly ordered structure in the thin film and the rest of the polymers follow in this order. It is also clear from the in-plane diffractions (Figure 4b) that PNDT3BT-16 forms the most ordered structure. PNDT3BT-16 exhibited a very strong (010) peak at $2\theta = 24.5°$ corresponding to the π-stacking order between facing backbones with a distance (d_π) of 3.6 Å, along with a small peak at $2\theta = 20.0°$ most likely assignable to (004), the backbone repeat unit. PNDT4BT-16 also showed a sharp (010) peak ($d_\pi = 3.6$ Å), but without a diffraction peak corresponding to the backbone repeat unit. It is noteworthy that this d_π of 3.6 Å for these two polymers are smaller than those of so far reported high-performance semiconducting polymers that typically possess 3.7–3.8 Å. These indicate that PNDT3BT has a more ordered backbone structure than PNDT4BT, though the π-stacking crystallinity seems to be comparable. On the other hand, in PNDT1BT-16 a diffraction corresponding to (010) was barely seen at around $2\theta = 24.0°$ (d_π = ca. 3.7 Å), indicative of very weak π-stacking crystallinity. In PNDT2BT-16, although some unassignable diffraction peaks appeared at $2\theta = 16$~$22°$, no peak was found at around $2\theta = 24$~$25°$, indicating that the π-stacking order is absent. The ordering structure is one of the most important parameter for the charge carrier transport, and it is mostly accepted that the higher mobility would be achieved as the ordering structure enhances. Hence, these results perfectly agree with the order of the mobility for these polymers as shown above.

IOP Conf. Series: Materials Science and Engineering **54** (2014) 012016 doi:10.1088/1757-899X/54/1/012016

3. Naphthobisthiadiazole-based polymer

Figure 5 shows chemical structure novel semiconducting polymers based on naphthobisthiadiazole (PNTz4T) and benzothiadiazole (PBTz4T) [12]. Naphthobisthiadiazole (NTz) is a doubly benzothiadiazole (BTz)-fused heterocycle [12-14]. Since NTz is expected to be a more rigid π–extended ring as compared to BTz, the incorporation into the polymer main chain should enhance intermolecular interaction and thereby promote a strong π–π stacking structure due to the more coplanar and the better arranged backbone structure compared to the BTz-analogue. In fact, the UV-vis absorption spectra of the polymer thin films show that PNTz4T gives a more defined structure, in which the absorption peak is more sharpened, than PBTz4T (Figure 6). The red-shift of the absorption range for PNTz4T than that for PBTz4T should be due to the stronger electron-accepting nature of the NTz moiety than BTz, which would lowers the level of the lowest occupied molecular orbital and thus the bandgap.

OFETs of the polymers are fabricated and evaluated using top-contact, bottom-gate devices using the polymer thin films spin-coated from *o*-dichlorobenzene (DCB) solutions, which are subsequently annealed at 200 °C. Figure 7a depicts the typical current–voltage curves of the transistor devices based on PNTz4T and PBTz4T. Mobilities evaluated from the saturation regime for PNTz4T are as high as 0.6 cm^2/Vs, and typically in a range of 0.2~0.4 cm^2/Vs with current on/off ratios of ~10^7. The mobility of PNTz4T is 1 order of magnitude higher than that of PBTz4T, typically 0.05 cm^2/Vs, indicating the high potential of the NTz core.

OPV cells are fabricated by spin coating the solution of the polymer blended with [6,6]-phenyl

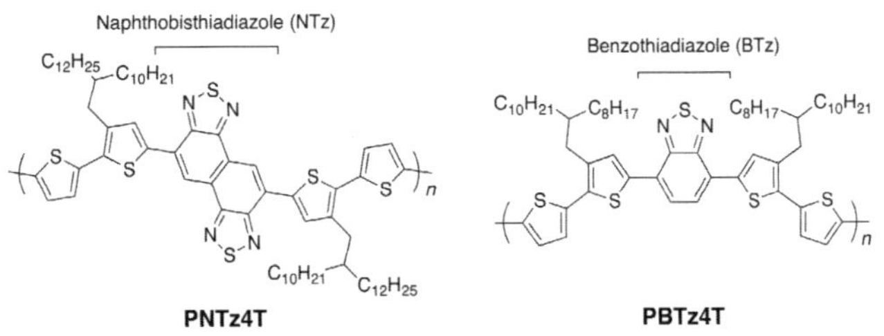

Figure 5. Chemical structures of PNTz4T and PBTz4T.

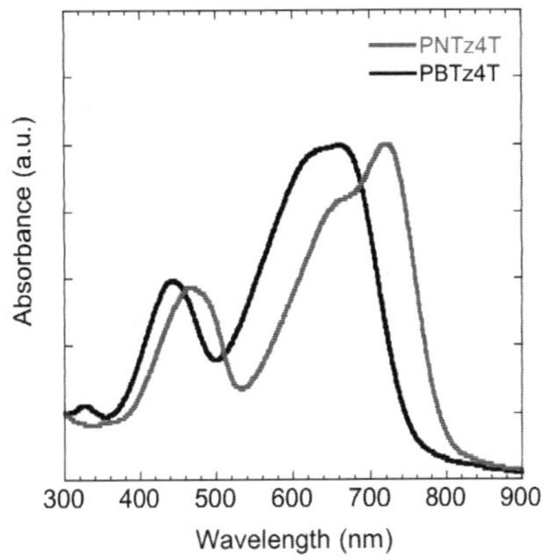

Figure 6. UV-vis absorption spectra of PNTz4T and PBTz4T in the thin film.

C61-butyric acid methyl ester ($PC_{61}BM$) in DCB. Current density–voltage curves of the devices under one sun of simulated AM 1.5G solar irradiation (100 mW/cm^2) are displayed in Figure 7b. While PBTz4T shows power conversion efficiencies (PCE) of ~2.6% with the polymer to $PC_{61}BM$ weight ratio (p:n) of 1:1, in which the short circuit current (J_{SC}) = 5.6 mA/cm^2, open-circuit voltage (V_{OC}) of 0.74 V, and fill factor (FF) of 0.63, PNTz4T shows a maximum PCE of 6.3% with J_{SC} = 12.0 mA/cm^2, V_{OC} = 0.76 V, and FF = 0.69 at p:n = 1:1.5. Significantly higher J_{SC} for the devices with PNTz4T than with PBTz4T is most likely as a result of, in part, the wider absorption range of PNTz4T.

In order to further understand these higher performances of PNTz4T in transistors and solar cells, the ordering structures of the polymers are investigated by the grazing incidence X-ray diffraction measurements with a two dimensional detector. The diffraction images of the polymer-only films and

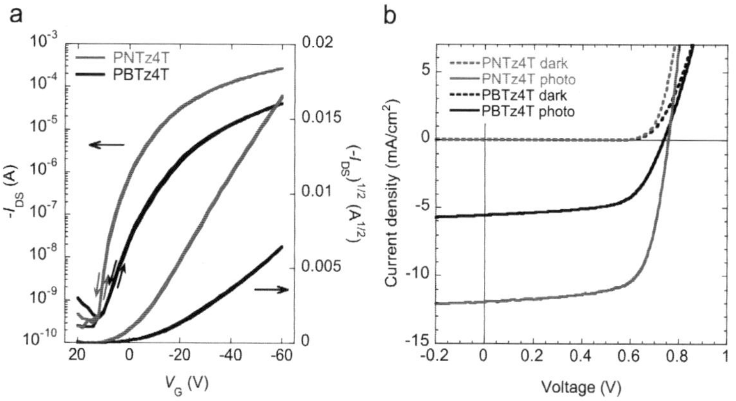

Figure 7. Current–voltage curves of the transistors (a) and solar cells (b) using the present polymers.

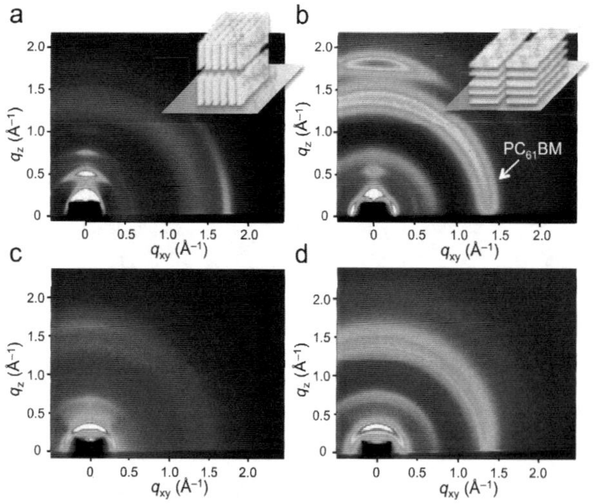

Figure 8. 2D-GIXD images of the PNTz4T thin film (a), PNTz4T/$PC_{61}BM$ blend film (b), PBTz4T thin film (c), and PBTz4T/$PC_{61}BM$ blend film (d). The illustrations shown in Figure 8a and 8b depict the motif of edge-on and face-on orientation, respectively.

polymer/$PC_{61}BM$ blend films are displayed in Figure 8, where they reflect the ordering structures in the transistors and solar cells, respectively. In the polymer-only film of PNTz4T (Figure 8a), diffractions assignable to the lamellar ($q_z \approx 0.25$ Å^{-1}) and the π–π stacking structures ($q_{xy} \approx 1.7$ Å^{-1}) appear on the q_z and q_{xy} axes, respectively, indicating the predominant edge-on orientation, i.e., the polymer backbones stand on the substrate surface [12]. In addition, the π–π stacking distance of PNTz4T is determined to be 3.5 Å, which is very narrow for semiconducting polymers, and thus this well rationalizes the high mobilities of PNTz4T in transistors. In contrast, PBTz4T shows largely arcing diffraction corresponding to the lamellar structure, indicating that there is no preferential orientation (Figure 8b). PBTz4T does not show clear diffraction corresponding to the π–π stacking structure, indicative of the much lesser crystalline nature of PBTz4T in the thin film. This large difference in the ordering structure between PNTz4T and PBTz4T is in good agreement with the fact that the transistor performances are quite distinct.

In the PNTz4T/$PC_{61}BM$ blend film, it is interesting that PNTz4T mainly orients in a face-on manner, i.e., the polymer backbones lie on the substrate, as the diffraction corresponding to the π–π stacking appears on the q_z axis (Figure 8c). It should also be noted that PNTz4T still has the narrow π–π stacking distance of 3.5 Å in the blend film. In the meantime, PBTz4T provides much less crystalline feature in the blend film, where there is no π–π stacking diffraction (Figure 8d). The predominant face-on orientation and the preserved narrow d_π of PNTz4T in the blend film should allow efficient charge transport in the cells, where the films are sandwiched between the electrode, which could be one of the main reasons that PNTz4T shows high PCE of exceeding 6% and greater performances as compared to PBTz4T.

4. Summary

In summary, structure–property relationships of a novel semiconducting polymer based on NDTs and NTz are presented. A striking structural feature of these polymers is the strong π–π stacking structure with a narrow distance of 3.5–3.6 Å, most likely owing to its π-extended rigid structure, allowing better arrangement of the main chain and the side chain. As a result, these polymers showed high mobilities in transistor devices, and, in particular, together with the broad absorption spectrum, PNTz4T provided high photovoltaic properties in solar cells.

References

[1] Facchetti A 2011 *Chem. Mater.* **23** 733
[2] Beaujuge P M, Fréchet, J M J 2011 *J. Am. Chem. Soc.* **133** 20009
[3] McCullough R D 1998 *Adv. Mater.* **10** 93
[4] Sirringhaus H, Brown P J, Friend R H, Nielsen M M, Bechgaard K, Langeveld-Voss B M W, Spiering A J H, Janssen R A J, Meijer E W, Herwig P, De Leeuw D M 1999 *Nature* **401** 685
[5] Osaka I, McCullough R D 2008 *Acc. Chem. Res.* **41** 1202
[6] McCulloch I, Heeney M, Chabinyc M L, DeLongchamp D, Kline R J, Colle M, Duffy W, Fischer D, Gundlach D, Hamadani B, Hamilton R, Richter L, Salleo A, Shkunov M, Sparrowe D, Tierney S, Zhang W 2009 *Adv. Mater.* **21** 1091
[7] Shinamura S, Miyazaki E, Takimiya K 2010 *J. Org. Chem.* **75** 1228
[8] Shinamura S, Osaka I, Miyazaki E, Nakao A, Yamagishi M, Takeya J, Takimiya K 2011 *J. Am. Chem. Soc.* **133** 5024
[9] Osaka I, Abe T, Shinamura S, Miyazaki E, Takimiya K 2010 *J. Am. Chem. Soc.* **132,** *132*, 5000
[10] Osaka I, Abe T, Shinamura S, Takimiya K 2011 *J. Am. Chem. Soc.* **133** 6852
[11] Osaka I, Shinamura S, Abe T, Takimiya K 2013 *J. Mater. Chem. C* **1** 1297
[12] Osaka I, Shimawaki M, Mori H, Doi I, Miyazaki E, Koganezawa T, Takimiya K 2012 *J. Am. Chem. Soc.* **134** 3498
[13] Wang M, Hu X, Liu P, Li W, Gong X, Huang F, Cao Y 2011 *J. Am. Chem. Soc.* **133** 9638
[14] Mataka S, Takahashi K, Ikezaki Y, Hatta T, Tori-i A, Tashiro M 1991 *Bull. Chem. Soc. Jpn.* **64** 68

IOP Conf. Series: Materials Science and Engineering **54** (2014) 012017 doi:10.1088/1757-899X/54/1/012017

Syntheses of light emitting poly(*N*-aryl-2,7-carbazole)s

Masashi Kijima[1,2]

[1]Division of Materials Science, Faculty of Pure and Applied Sciences, University of Tsukuba, Tennodai 1-1-1, Tsukuba, Ibaraki 305-8573, Japan
[2]Tsukuba Research Center for Interdisciplinary Materials Science (TIMS)

E-mail: kijima@ims.tsukuba.ac.jp

Abstract. Recent remarkable development of poly(2,7-carbazole)s and copolymers including a carbazolyl unit have demonstrated that they are one of key conjugated polymer materials for optoelectronic applications. This short review reports recent progress made in synthesis and characterization of poly(*N*-aryl-2,7-carbazole)s for application of light emitting diode. Main strategy and remaining challenges in the development of reliable emitting materials for devices of organic light emitting diodes are discussed.

1. Introduction

For many years, pyrrole-based materials have been investigated for the purpose of their electrical and optoelectronic applications, but there were few researches on light emissive applications except polycarbazoles. However, poly(3,6-carbazole)s have not been drawn much attention as emissive materials for organic light emitting diode (OLED), because they did not have excellent properties, *i.e.*, low processability to make uniform thin films due to not so-high average molecular weights, low fluorescence quantum efficiency, and wide emission spectra in the blue light region. Initially, several polyindoles were synthesized and investigated to develop novel pyrrole-based blue emitting materials for availability of monomers and easy modification at N-position [1]. They were superior to poly(3,6-carbazole)s in terms of fluorescence (PL) quantum efficiency, color purity of blue PL in the solid state, and processability for thin films. However, OLED devices using poly(1-dodecyl-2,6-indolylene-1,4-phenylene) and poly(1-(4-dodecyloxyphenyl)-3-methyl-2,6-indolylene) as an emitting layer material only showed weak bluish emissions less than 100 cd/m^2 with broad emission spectra (λ_{max} = 470~480 nm), when they energized at about 20 V. Around the same time, Morin and Leclerc have successfully synthesized poly(*N*-alkyl-2,7-carbazole)s [2], which were certainly not only a potential candidate for blue light emitting materials for OLED but also capable materials for other semiconducting applications such as organic solar cells [3].

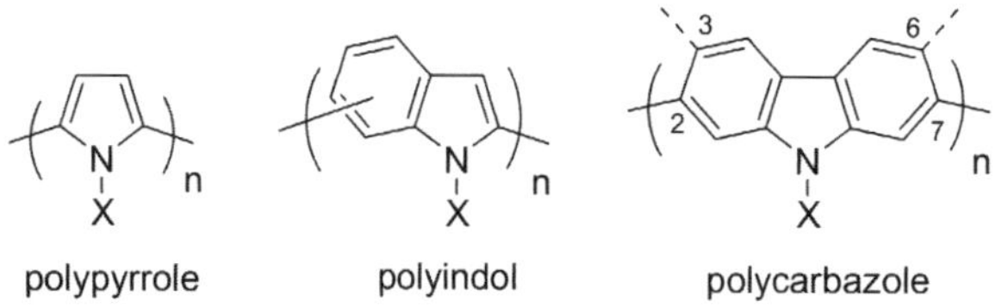

Figure 1. Structures of pyrrole-based conjugated polymers.

2. Poly(*N*-aryl-2,7-carbazole)s

2.1. Synthesis

Several poly(*N*-aryl-2,7-carbazole)s, whose N-substituent is an aryl group instead of alkyl, have been synthesized by our group [4-9] and others [10-12] according to Scheme 1. N-Arylation of 2,7-dichlorocarbazole (X= Cl) [2] with halogenobenzene derivatives (Ar-Y) was carried out by CuI-*trans*-1,2-cyclohexanediamine catalyzed method [13]. In the case of arylation of 2,7-dibromocarbazole (X= Br) [14] using this method, reaction conditions to suppress formation of by-products such as carbazolylcarbazoles must be chosen. Reactions of *9H*-carbazoles with fluoronitrobenzene (Y= F) under basic conditions are a good selective method for introducing an aniline moiety at N-position [7, 10, 15]. Usually poly(*N*-aryl-2,7-carbazole)s can be obtained by Yamamoto dehalogenative condensations of 2,7-dihalogenocarbazoles using a zerovalent Ni complex or by Suzuki coupling reaction of 2,7-dihalogenocarbazole with 2,7-diboronated carbazole using a Pd catalyst. Chemical and photophysical properties of typical polymers, **P1**, **P2**, and **P3**, synthesized from corresponding dihalogenomonomers (**1**, **2** and **3**) by the Yamamoto method, are discussed in the following sections.

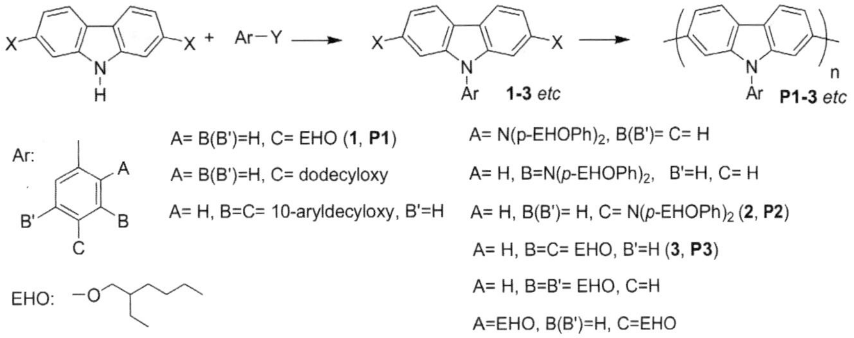

Scheme 1. Reaction pathway for synthesis of monomers and polymers.

2.2. General properties

Poly(*N*-aryl-2,7-carbazole)s having alkyl chains such as 2-ethylhexyloxy group (EHO) at N-aryl moiety shown in Scheme 1 were sufficiently soluble in organic solvents such as THF and $CHCl_3$. GPC analysis showed number average molecular weights of these polymers were more than 5 kDa and generally around 10~30 kDa. They also showed a good thermal stability with a high glass transition temperature (T_g > 150°C), a high softening temperature (T_s > 210°C), and high degradation temperatures above T_d = 350°C, at which temperature 5% of mass is reduced from the starting material in the thermal gravimetric analysis. They absorbed light around 380—390 nm (λ_{max}) and fluoresced around 410—420 nm (λ_{max}) in $CHCl_3$ with high quantum efficiency about 0.8 in $CHCl_3$.

2.3. OLED characteristics

A typical sample is poly(*N*-(2-ethylhexyloxy-4-phenyl)-2,7-carbazole) **P1** [4, 5], which showed the photoabsorption λ_{max}= 381 nm in film state, PL emission λ_{max}= 432 nm in film state, almost quantitative PL quantum yield in $CHCl_3$, wide band gap (E_g) between highest occupied molecular orbital (HOMO) and lowest unoccupied molecular orbital (LUMO) about 2.90 eV, and an energy level of HOMO about −5.45 eV (Figure 2). The deeper energy level of HOMO and higher quantum efficiency of **P1** than corresponding N-alkyl substituted ones are advantageous to be applied poly(*N*-aryl-2,7-carbazole)s in OLED. These energy levels and E_g can be tuned by changing the N-aryl group of the polymers [7]; *i.e.*, the HOMO level of **P2** having an electron donating triphenylamino group at *m*-position was shallower (−5.27 eV) than that of **P1**, while that of **P3** having an alkoxy group at *m*-position was deeper (−5.62 eV).

The OLED device having a configuration of ITO/poly(3,4-ethylenedioxythiophene doped)-poly(styrenesulfonic acid) (PEDOT-PSS) (100 nm)/Polymer **P1** (100 nm)/Ca+Al (300 nm) showed blue emission started from a low turn on voltage (3 V), a maximum luminance (L_{max}) (about 1500 cd/m^2), and a current efficiency η_c higher than 0.1 cd/A at 10 V [4]. Other polymers (**P2** and **P3**) having similar configuration of the device; ITO (150 nm)/PEDOT:PSS (60 nm)/polymer/Ba (5 nm)/Al (80 nm), showed EL spectra similar to PL with brightness over 1000 cd/m^2 at 10 V, which suggested that poly(*N*-aryl-2,7-carbazole)s were a potential material for blue OLED [8].

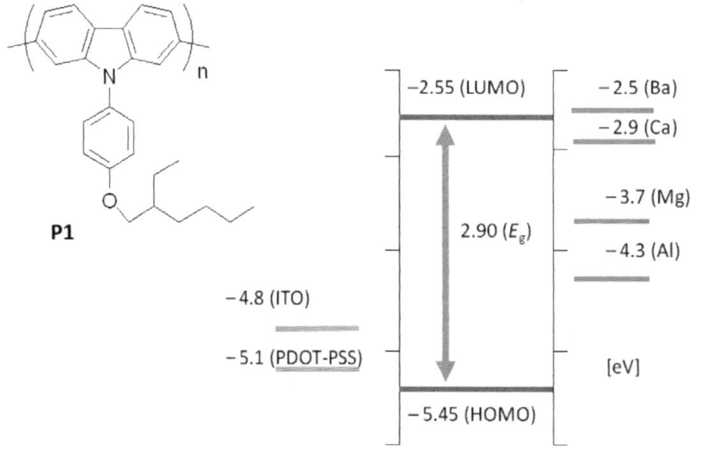

Figure 2. Chemical structure and energy diagram of poly(*N*-(2-ethylhexyloxy-4-phenyl)-2,7-carbazole) **1** with reference data of work functions of typical materials.

Figure 3 summarizes OLED characteristics of a device, ITO (150 nm)/PEDOT-PSS (90 nm)/**P1** (70 nm)/CsF (2 nm)/Al (120 nm) [5]. It began to emit at 4 V, and brightness seemed to increase monotonically as increasing applied voltages to reach a L_{max} of 34600 cd/m^2 at 10 V with a high current efficiency (η_c) (1.5 cd/A), and after that, luminance sharply decreased above 11 V but current continued to flow as shown in Figure 3a. These results suggest that the device damages during operations. A turning point can be recognized at 7 V as shown in Figure 3b in the changes of power and current efficiencies (η_p, η_c) and international commission on illumination (CIE) chromaticity coordinates (x, y) *vs*. operating voltages. The emitting color of the device was changed from blue (0.18, 0.15) at 7 V to blue green (0.22, 0.32) at 9 V before and after the turning point, then it changed green (0.24, 0.42) above 10 V in the last result. These color changes corresponding to spectra change of electroluminescence (EL) are summarized in Figure 4. The EL spectra under operating voltages below 7 V were similar to the PL spectrum, but EL spectra observed above 9 V were substantially different. This implies that different exciton might be predominantly formed under higher voltages than the turning voltage. In addition, each blue EL emission spectrum and luminance *vs*. operating voltages between 4-7 V were reversible and reproducible for several scans, but applying voltages over 9 V annihilated the reversibility within a few scans.

The low turn on voltages (about 4 V) but the low current efficiency observed in this device suggests that the energy barrier between the electrodes, interlayer, and the polymer emitting layer might not be serious but balanced exciton formation which could be controlled by carriers' injection and transport must be inappropriate. In general, unbalanced operations affect stability of electroluminescence (EL) color and luminance of the OLED devices.

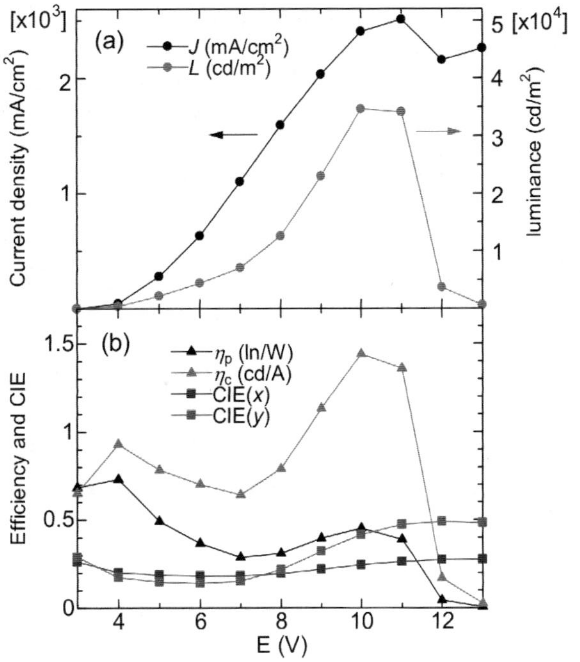

Figure 3. Current density and luminance (a) and efficiencies and CIE chromaticity coordinates (b) *vs*. operating voltages in the OLED device using **P1** as the emitting layer material.

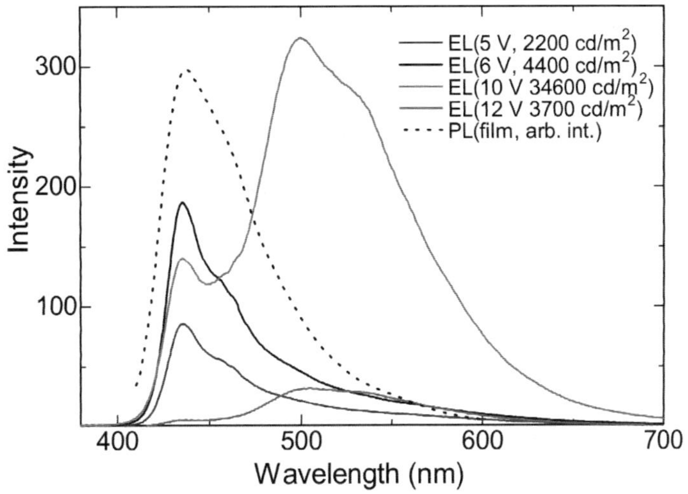

Figure 4. Electroluminescence spectra of the OLED device using **P1** as emitter at various operating voltages and the reference PL spectrum in film state.

2.4 Modifications of poly(2,7-carbazole)s for OLED.

Stability of emission color during operation of OLED devices is intimately related to whether an exciton dominantly generated by balanced carriers' injection and transport. Since polycarbazoles are good hole injecting and transport materials, the higher operating voltages are applied in the device, the more holes are accumulated in the emitting layer. The supersaturated state of holes in the polymer must cause irreversible reactions at reactive 3,6-CH of carbazole [16] to form another π-extended low energy states or formation of electroplexes [17]. In order to suppress these unfavorable emissions, three attempts have been done by i) introduction of a buffering group for hole injection/transport, ii) copolymerization with electron accepting unit for carriers' balance injection/transport, and iii) protection of 3,5-positions of poly(2,7-carbazole)s to stabilize them against electrooxidation.

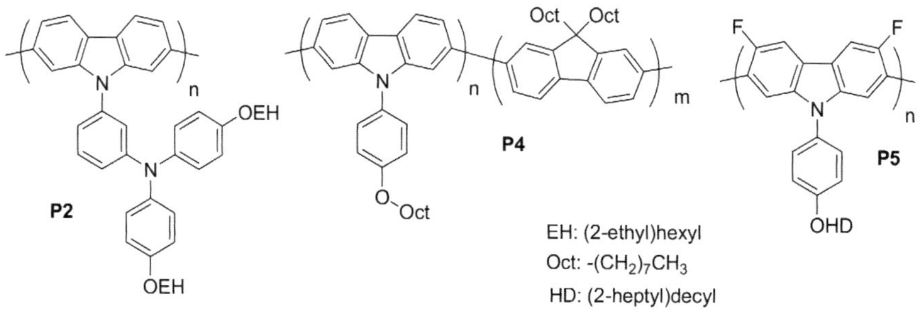

Figure 5. Chemical structures of typical functionalized poly(*N*-aryl-2,7-carbazole)s.

2.4.1. Buffering effect. Since polymers with triphenylamine units have been known as useful materials for hole injection and transport [18], triphenylamine moieties in poly(*N*-aryl-2,7-carbazole)s could function to buffer holes, which might suppress color change due to unfavorable oxidation of the conjugated main chain during device operations. **P2** having a triphenylamine moiety at the N-position (Figure 5) was also good blue emitting polymer, which showed absorption λ_{max}= 391 nm in $CHCl_3$, absorption λ_{max}=397 nm in film state (E_g = 2.9 eV), PL λ_{max} = 418 nm in $CHCl_3$ (ϕ_{PL}= 0.32) PL λ_{max} =450 nm in the thin film state, and the HOMO level= −5.27 eV. Comparing the performance between **P2** and **P3** in the OLED device consisting of ITO/PEDOT:PSS/polymer/Ba/Al, color purity of blue and other performances for the **P2**-device (EL λ_{max}= 450 nm, L= 1250 cd/m^2, CIE(x= 0.16, y= 0.17) at 10 V, η_c= 0.52 cd/A) were superior to those for the **P3**-device (EL λ_{max}= 484 nm, L= 1090 cd/m^2, CIE(x= 0.20, y= 0.24) at 10 V, η_c= 0.18 cd/A) [8]. A different OLED (ITO/PEDOT:PSS/polymer/ CsF/Al) device embedded a polymer similar to **P2** showed similar performances; EL λ_{max}= 426 nm, CIE (x= 0.18, y= 0.14), and 710 cd/m^2 at 9 V (η_c= 0.20 cd/A) [9]. This polymer also showed a turning point at 9 V as well as **P1**, but color change was moderate from purplish blue (0.18, 0.12) at 6 V to blue green (0.21, 3.0) at 15 V via blue (0.19, 0.20) at 10 V (L_{max}= 1170 cd/m^2). These results suggest that the triphenylamine structure at N-position of poly(*N*-aryl-2,7-carbazole)s is effective to realize and sustain blue emission under driving voltages lower than 10 V.

2.4.2. Copolymerization effect. If electron injection ability of the polycarbazoles is improved by lowering the level of LUMO, the EL luminance and efficiency will increase and the EL color change could be suppressed. Copolymerization of the carbazolyl unit with other ones might be a good strategy to change the HOMO and LUMO energy levels, which should attain balanced carriers' injection/transport in poly(*N*-aryl-2,7-carbazole) based polymers. It has been already attempted that *N*-alkylcarbazole-2,7-diyl and fluorene-2,7-diyl based copolycondensates were synthesized to decrease barrier height for hole injection for the purpose of heightening the OLED performances of polyfluorenes [19,20]. From another perspective, a random copolycondensate, **P4** (Figure 5), was prepared by the Ni-mediated polycondensation of 2,7-dibromo-*N*-(octyloxyphenyl)carbazole and 2,7-dibromo-9,9-dioctylfluorene) in molar ratio of 1:1 [6] to lower the LUMO energy level of polycarbazole. The copolycondensate **P4** absorbed light energy at about λ_{max}= 390 nm in $CHCl_3$ and 400 nm in film state (E_g = 2.9 eV), and fluoresced at about λ_{max} = 417 nm in $CHCl_3$ (ϕ_{PL}= 0.90) and 430 nm in the thin film state, which were almost same to those of **P1**. The energy level of HOMO and LUMO situated lower than those of **P1** and higher than those of poly(octylfluorene). The OLED device having the configuration of ITO/PEDOT:PSS/**P4**/CsF/Al showed bluish EL (CIE: x= 0.21, y= 0.22) at operating voltages lower than 7 V, and notably higher performance, L_{max} of 31200 cd/m^2 at 11.0 V. The current efficiencies (*e.g.*, $\eta_{c\ max}$=1.34 cd/A [at 4 V]) observed under operating voltages lower than 7 V were higher than those of the other devices using polycarbazole reported so far. The color purity of blue emissions observed under operating voltages lower than 7 V was inferior to that of **P1**, but gradual color change from light blue (around the turn on voltage) to bluish green ((x= 0.23, y= 0.37) at 12 V) during the operation was observed, which was comparable to polyfluorene. From these results, it is concluded that stability of the copolymer **P4** in the device is situated between **P1** and polyfluorene, but current efficiency of the **P4**-device is superior to the devices embedded homopolymers under operating voltages below the turning point (7 V), which is considered to be due to improved balance of carriers' injection and transport.

2.4.3. Protection effect. Since presence of reactive C-H at 3,6-positions of carbazole is also considered as one of the factors of the color change during device operations, various 3,6-substututed poly(2,7-carbazole)s were designed, synthesized, and characterized in order to enhance EL efficiency and stability of the emission color. Poly(3,6-dimethyl-9-alkyl-9*H*-carbazole-2,7-diyl)s and related alternating copolymer have been synthesized and characterized [21]. The 3,6-dimethyl substituted homopolymer showed an E_g about 3.2 eV, HOMO level= −5.5 eV, fluorescence λ_{max}= 397 nm in the film state, and rather low ϕ_{PL}=0.16 in CH_2Cl_2. These results suggests that steric hindrance of the 3,6-dimethyl substituents reduces π-conjugation along the main chain of poly(2,7-carbazole). Similarly, poly(3,6-difluoro-9-(4-(2-butyloctyloxy)phenyl)-9*H*-carbazole-2,7-diyl) showed blue fluorescence (λ_{max} = 436 nm) in the film state, ϕ_{PL}= 0.27 in solution, E_g= 3.07 eV, and a deeper HOMO level = −5.83 eV [11], which showed better photophysical properties than the methyl derivative to examine OLED behavior. Thus, **P5** shown in Figure 5 was used as a typical example to examine the effect of 3,6-substituents in poly(2,7-carbazole)s. **P5** showed blue fluorescence (λ_{max} = 431 nm) in the film state, ϕ_{PL}= 0.31 in $CHCl_3$, E_g= 3.08 eV, and HOMO level= −5.77 eV [22], which were almost same to the above data of poly(3,6-difluoro-*N*-arylarbazole) [11]. The OLED device having the configuration of ITO/PEDOT:PSS/**P5**/CsF/Al showed blue EL at operating voltages lower than 10 V (CIE: x= 0.17, y= 0.09), $\eta_{c\ max}$=0.33 cd/A [at 5.5 V], and L_{max} of 490 cd /m^2 at 8.0 V. The luminance was not so high but an ideal deep blue color near the primary blue of the national television system committee (NTSC) in1953 was attained. The deep blue emission was stable under operating voltages below 10 V, and the color change from blue to greenish blue gradually occurred over 10 V. The good color purity and stability is ascribed to the 3,6-difluoro protection, whose steric and electronic effects enhance the chemical stability, slightly widen the E_g, deepen the HOMO level, and suppress π-stacking.

3. Conclusions

Several poly(*N*-ary-2,7-carbazple)s were synthesized, characterized, and evaluated as a blue emitting material for OLED. Basically, poly(*N*-ary-2,7-carbazole)s were good emitting materials for blue photoluminescence in solution and films with good quantum efficiency. The HOMO level of the polymers could be controlled from −5.3 eV to −5.8 eV by modification of the N-aryl moiety. Several OLED devices emitting blue light could be made by using the poly(*N*-ary-2,7-carbazole)s as the emitting layer material. A problem was the irreversible change of the emitting color from blue to green during operation under applied voltages higher than a turning point. It is assumed that carriers' injections from electrodes, transportations, and recombination to form an exciton are in counterpoise

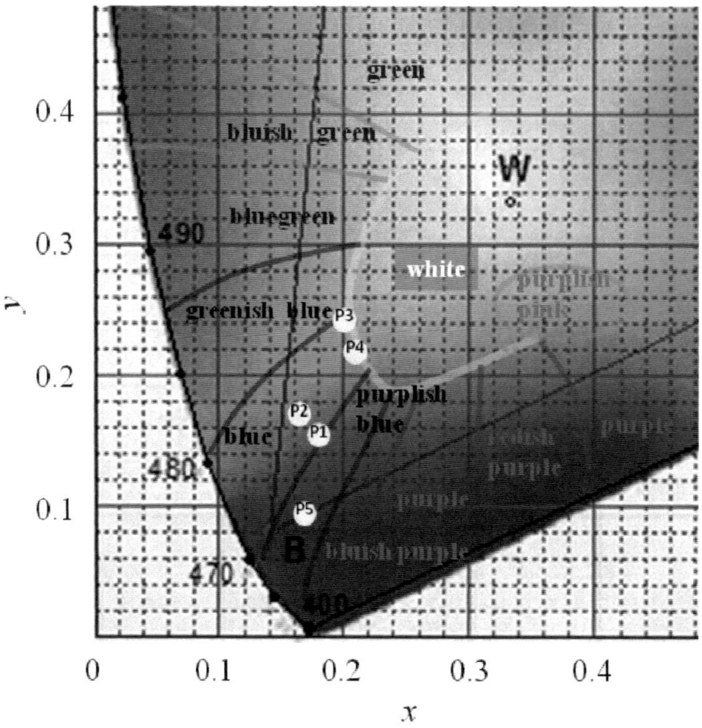

Figure 6. Blue emission colors (CIE (*x*, *y*)) of OLED devices embedded **P1** (7 V), **P2** (10 V), **P3** (10 V), **P4** (7 V), and **P5** (8 V) on the 1931 CIE chromaticity diagram with reference points of NTSC primary blue in1953: B (0.14, 0.08) and achromatic point: W (0.33, 0.33).

under operating voltages below the turning point, while undesired oxidations of the polymer or emissions from other excitons such as electroplex under supersaturated state of holes occur above the turning voltage. In order to increase EL efficiency and suppress the unfavorable emissions, balanced carriers' injection and transport in the OLED were attempted by i) introduction of a buffering group for hole injection and transport, ii) copolymerization with electron accepting unit to stabilize against oxidation and to control balance of carriers' injection and transport, and iii) protection of 3,5-positions of poly(2,7-carbazole)s to prevent from unfavorable electrooxidations. Figure 6 indicates CIE coordinates of typical blue light emitting poly(*N*-aryl-2,7-carbazole) based OLED devices, and Table 1 summarizes CIE (*x*, *y*) results of these investigations.

Poly(*N*-(2-ethylhexyloxy-4-phenyl)-2,7-carbazole) **P1**, a standard polymer, emitted blue (0.18, 0.15) below 7 V, but the change of color to green (*e.g.*, (0.24, 0.42) at 10 V) was not a little under the higher voltages. Another standard polymer, **P3**, which have two 2-ethylhexyloxy substituents at *m*, *p*-positions of the *N*-phenyl also behaved as well as **P1**. **P2** having a triphenylamine moiety at N-position kept to emit pure blue (0.16, 0.17) by 10 V compared to the color of **P3** (0.20, 0.24) at 10 V. This result suggests that the triphenylamine moiety in **P2** could function to buffer holes to suppress unfavorable oxidations. The copolymerization was also an effective manner to stabilize the emission color and heighten the EL efficiency. Practically, the initial emitting color of light blue (0.21, 0.22) for **P4** was kept under operating voltages below 7 V, furthermore, the efficiencies (η_c) were higher than those of respective homopolymers under these conditions. Nevertheless, the blue color gradually changed to bluish green (0.23, 0.37) at 12 V and at last green (0.27, 0.48) at 14 V. The protection of

3,6-positions with fluorine was effective to keep color purity of deep blue emission (0.17, 0.09) by 8 V. The color change according to varying applied voltages from 0 to 15 V was within range of blue, that is, the purplish blue color gradually changed to greenish blue (0.19, 0. 25) at 15 V via blue (0.18, 0.17) at 12 V. However, the luminance (L_{max}= 490 cd/m^2) and efficiency ($\eta_{c,\,max}$= 0.33) were relatively low, which suggests that carriers' injection and transport are unbalanced in the OLED device.

In conclusion, it is believed that further modifications with proper molecular design could develop the more stable and highly efficient blue light emitting poly(*N*-ary-2,7-carbazole)s for OLED.

Table 1. Summary of results on characteristics of blue light emitting and color change of poly(*N*-aryl-2,7-carbazole)-based OLED devices.

		luminance		CIE		
Polymer	Device[a]	L (cd m^{-2})	[at V]	(x, y)	Emitting color	Reference
P1	A	1500	[10][b]	(0.19, 0.16)[c]	blue (purplish)	4
	B	7080	[7]	(0.18, 0.15)	blue	6
	B	34600	[10][b]	(0.24, 0.42)	light green	6
P2	B	710	[9]	(0.18, 0.14)	blue (purplish)	9
	B	1170	[10][b]	(0.19, 0.20)	blue	9
	C	1250	[10][b]	(0.16, 0.17)[c]	blue	8
P3	B	5000	[8]	(0.17, 0.14)	blue	9
	B	18800	[11][b]	(0.22, 0.36)	light green (bluish)	9
	C	1090	[10]	(0.20, 0.24)[c]	pale blue (greenish)	8
P4	B	6400	[7]	(0.21, 0.22)[c]	pale blue	6
	B	31200	[11][b]	(0.23, 0.37)	light green (bluish)	6
P5	B	490	[8][b]	(0.17, 0.09)[c]	deep blue (purplish)	22

[a] Configuration of OLED devices: ITO/PEDOT:PSS/polymer/cathode (A = Ca/Al , B= CsF/Al, C= Ba/Al) .
[b] Maximum luminance.
[c] The CIE values were pointed in Figure 6.

Acknowledgement
The author thank Mr. T. Shinnai, Mr. K. Tsuchiya, Kanto Chemical Co. Inc. for device fabrication and evaluation of OLED, Prof. H. Goto, University of Tsukuba, for the measurement of GPC, and Chemical Analysis Division, Research Facility Center for Science and Technology, University of Tsukuba, for facilities of the NMR, elemental analysis, DSC, UV−vis, and PL measurements.

References

[1] Abe S, Nagasaki Y, and Kijima M 2003 *Synth. Met.* **135-136** 123
[2] Morin J-F and Leclerc M 2001 *Macromolecules* **34** 4680
[3] Blouin N and Leclerc M 2008 *Acc. Chem. Res.* **41** 1110
[4] Kijima M, Koguchi R, and Abe S 2005 *Chem. Lett.* **34** 900
[5] Mori T, Shinnai T, and Kijima M 2011 *Polym. Chem.* **2** 2830
[6] Horii T, Shinnai T, Tsuchiya K, Mori T, and Kijima M 2012 *J. Polym. Sci., Polym. Chem.* **50** 4557
[7] Kobayashi N, Koguchi R, and Kijima M 2006 *Macromolecules* **39** 9102
[8] Kobayashi N and Kijima M 2007 *App. Phys. Lett* **91** 081113.
[9] Koguchi R, Kobayashi N, Shinnai T, Oikawa K, Tsuchiya K, and Kijima M 2008 *Macromol.*

Chem. Phys. **209** 439
[10] Iraqi A, Simmance T G, Yi H, Stevenson M, and Lidzey D G 2006 *Chem. Mater.* **18** 5789
[11] Chauhan S S, Yi H, and Iraqi A 2008 *Mol. Cryst. Liq. Cryst.* **497** 129
[12] Wang L, Fu Y, Zhu L, Cui G, Liang F, Guo L, Zhang X, Xie Z, and Su Z 2011 *Polymer* **52** 1748
[13] Klapars A, Antilla J C, Huang X, and Buchwald S L 2001 *J. Am. Chem. Soc.* **123** 7727.
[14] Dierschke F, Grimsdale A C, and Müllen K 2003 *Synthesis* **16** 2470
[15] Jian H and Tour J M 2003 *J. Org. Chem.* **68** 5091
[16] Macit H, Sen S, Saçak M 2005 *J. Appl. Polym. Sci.* **96** 894
[17] Hayer A, Regemorter T V, Höfer B, Mak C S K, Beljonne D, Köhler A 2012 *J Polym Sci B Polym Phys* **50** 361
[18] Iwan A and Sek D 2011 Prog. Polym. Sci. **36** 1277
[19] Triapattur S, Belletête M, Drolet N, Leclerc M, and Durocher G 2003 *Chem. Phys. Lett.* **370** 799
[20] Drolet N, Beaupré S, Morin J F, Tao Y, and Leclerc M 2002 *J. Opt. A Pure Appl. Opt.* **4** S252
[21] Iraqi A, Pickup D F, and Yi H 2006 *Chem. Mater.* **18** 1007
[22] Ohashi T and Kijima M. 2013 *Polym. Prep. Jpn. (SPSJ)* **62** 3887

Elastic anomaly and order-disorder nature of multiferroic barium sodium niobate studied by broadband brillouin scattering

Shiori Ota, Kazuya Matsumoto, Kohei Suzuki, and Seiji Kojima*

Division of Materials Science, University of Tsukuba, Tsukuba, Ibaraki 305-8573, Japan

*E-mail: kojima@bk.tsukuba.ac.jp

Abstract. The successive phase transitions of multiferroic barium sodium niobate, $Ba_2NaNb_5O_{15}$ (BNN), were studied by Brillouin scattering. The LA, TA modes, and central peak were measured in a large temperature range from room temperature up to 750 °C. In the vicinity of a ferroelectric phase transition at about T_C = 585 °C from the prototypic tetragonal 4/mmm to ferroelectric 4mm phases, elastic anomaly was observed for LA and TA modes. In addition, the order-disorder nature was observed by the temperature dependence of a central peak. For further cooling another elastic anomaly was also observed in the vicinity of a ferroelastic incommensurate phase transition at about T_{IC} = 285 °C into orthorhombic 2mm phase with the appearance of incommensurate modulation. The large thermal hysteresis of elastic anomaly near T_{IC} can be attributed the typical feature of the type III incommensurate phase transition predicted recently by Ishibashi and Iwata (2013 *J. Phys. Soc. Jpn.* **82** 044703).

1. Introduction

Ferroic crystals have two or more orientation states without external magnetic, electric and stress fields, and can change from one to another state by the application of external fields [1]. Here any two of orientation states are identical or enatiomorphic in crystal structure, while they are different with the direction of arrangements of atoms or spins. Ferromagnetic, ferroelectric, and ferroelastic crystals possess a spontaneous magnetization, spontaneous polarization, and spontaneous strain, respectively. The term 'multiferroic' was originally understood as the simultaneous presence of two or all three primary ferroic properties in the same phase, i.e., ferromagnetism, ferroelectricity and ferroelasticity [2]. Barium sodium niobate $Ba_2NaNb_5O_{15}$ (BNN) with tungsten bronze structure (figure 1) is ferroelectric and ferroelastic at room temperature [3,4]. Multiferroic BNN undergoes successive phase transitions on cooling from a prototypic paraelectric and paraelastic tetragonal (crystal symmetry : *4/mmm*) phase. About 560 °C it transforms into a ferroelectric tetragonal *4mm* phase with a spontaneous polarization along the *c*-axis (ferroelectric phase transition), and about 300 °C, it transforms into a ferroelastic orthorhombic *2mm* phase with a spontaneous strain e_6. For further cooling the reentrant phase transition into a ferroelectric paraelastic phase with tetragonal *4mm* occurs at about -168 °C. It was also reported that the lower ferroelastic to paraelastic phase transition temperature increases under high pressure, and at room temperature it occurs at about 2.2 GPa [5,6]. Moreover, in BNN, the incommensurability measured from the point [1/4, 1/4, (1/2)] in the Brillouin zone is known to remain down to very low temperatures. Very recently this incommensurate phase transition of BNN is assigned to the new type III incommensurate phase, which is stabilized by the freezing of the soft mode belonging the lower branch induced by the parabolic splitting [7].

Figure 1 shows the tungsten-bronze structure of BNN. The A_1 and A_2 sites are occupied by Na and Ba, respectively. The B_1 and B_2 sites are occupied by Nb, and the C_1 sites are vacancies. The *a*-axis and *b*-axis are rotated 45 degrees when the structure changes from tetragonal to orthorhombic system. There is no disorder of cations and vacancy in A_1 and A_2 sites, and BNN undergoes a normal ferroelectric phase transition in contrast to typical uniaxial relaxor $Sr_xBa_{1-x}Nb_2O_6$ in which random quenched fields appears owing to the disorder in A_1 and A_2 sites. As to the mechanism of ferroelectricity, up to the present there is no report of soft optic mode observed by Raman scattering, inelastic neutron and X-ray scattering experiments. The dispersion in the GHz range, which gives the possibility of order-disorder nature, was suggested by ultrasonic study [8]. Nevertheless, there is still no report of critical slowing down near the Curie temperature. Regarding the elastic anomaly, the temperature dependence of elastic properties of BNN was recently studied by the resonant ultrasonic spectroscopy (RUS), and the lock-in transition was detected at -233 °C [9]. However in RUS, it is difficult to measure elastic properties when the acoustic absorption increases markedly near a phase transition temperature or ferroic multi-domains state appears. While in Brillouin scattering it is possible to measure even if the remarkable increase of absorption occurs. In the present work order-disorder nature and elastic anomaly in the GHz frequency range were studied by Brillouin scattering in the large temperature range.

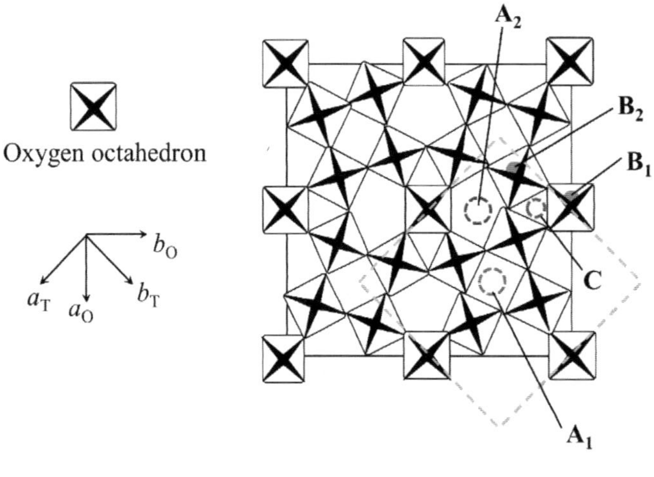

Figure 1. Tungsten-bronze-type structure on the (001) plane. The orthorhombic cell and the tetragonal cell are shown by solid and dotted lines, respectively.

2. Experimental

The colorless and transparent BNN crystal studied was grown by Czochralski method in Tamagawa factory, NEC. The (110) plate with the size of $2.65 \times 2.25 \times 0.68$ mm with two optically polished surfaces was used for Brillouin scattering measurement. Brillouin scattering spectra were measured at the backward scattering geometry using a tandem Fabry-Perot interferometer and a conventional photon counting system. A single frequency green YAG laser ($\lambda = 532$ nm) with power of 100 mw was used as an exciting source. The light spot size at a sample surface was about 10 μm using the optical microscope (BX-60) [10]. The temperature of a sample was controlled by the heating stage of a T1500 (high T-Linkam) from room temperature up to 750 °C. All Brillouin scattering spectra were measured in the condition that the free spectral range (FSR) and the scan range are 75 GHz, ± 65 GHz, respectively.

3. Results and Discussion

3.1. Brillouin scattering

By measuring Brillouin scattering, we can study sound velocity of acoustic attenuation of longitudinal acoustic (LA) and transverse acoustic (TA) modes and the relaxation process in the GHz range. The LA and TA sound velocity V_L and V_T, and their acoustic attenuation coefficients α_L and α_T were calculated from the Brillouin shift ν_B and the width Γ_B of LA and TA modes, respectively, by the eqs. (1) and (2),

$$V_i = \lambda_0 \nu_{Bi} / 2n\sin(\theta/2), \qquad (1)$$

and

$$\alpha_i = \pi\Gamma_{Bi}/V_i, \qquad (2)$$

where i denotes LA or TA, n, λ_0, θ is refractive index, the wavelength of incident beam (532 nm), scattering angle (180°), respectively. As to the relaxation process of polarization fluctuations in the GHz frequency range, we can determine the relaxation time τ by the equation,

$$\tau = 1/\pi\Gamma_{CP}, \qquad (3)$$

where Γ_{CP} is the width of a central peak. Therefore, in the study of a phase transition Brillouin scattering is the powerful tool to investigate not only elastic anomaly using LA and TA peaks but also relaxation process related to a central peak [10].

3.2. Ferroelectric phase transition

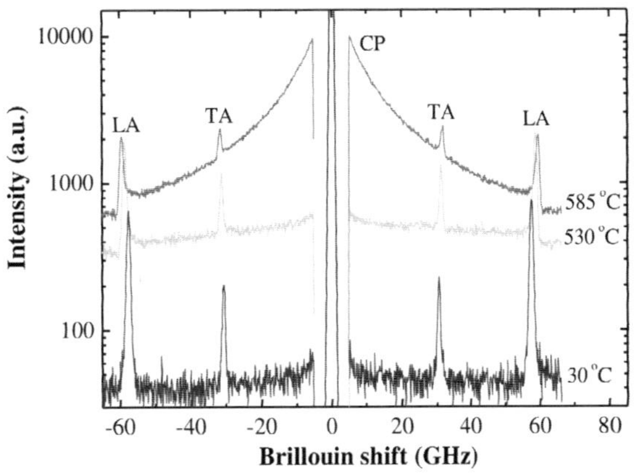

Figure 2. Brillouin scattering spectra of BNN. CP, LA, and TA denote a central peak, a longitudinal acoustic mode, and a transverse acoustic mode, respectively.

Figure 2 shows the temperature dependence of Brillouin scattering spectra. LA and TA modes are clearly observed as sharp peaks at about 58 and 30 GHz, respectively. In addition a central peak (CP) is observed at zero frequency shift near the ferroelectric phase transition temperature, T_C = 585 °C. The CP appears gradually below T_C, and its intensity shows the maximum at T_C.

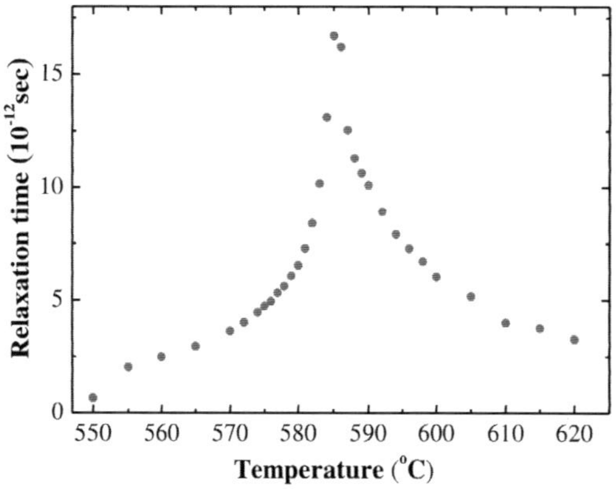

Figure 3. The temperature dependence of the relaxation time of BNN determined by the width of a central peak on heating.

Figure 3 shows the temperature dependence of relaxation time. The critical slowing down is clearly observed near T_C. According the IR study, no soft mode was observed down to 15 cm^{-1}, while the relaxation mode was not observed in their far IR measurement [11]. This result of figure 3 is the clear evidence to confirm the fact that the ferroelectricity of BNN is not displacive type but order-disorder type.

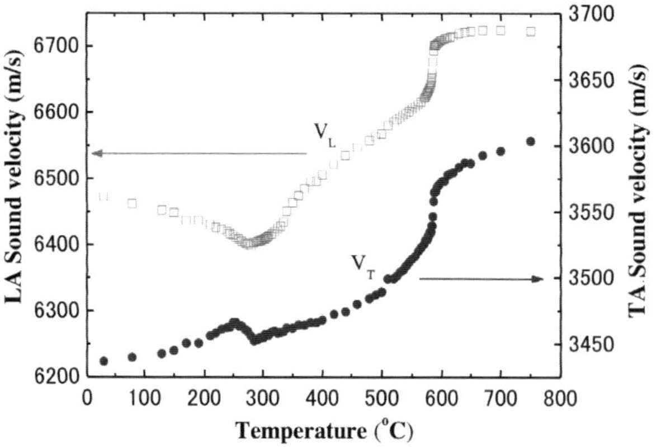

Figure 4. Sound velocity of longitudinal acoustic (LA) mode and transverse acoustic (TA) mode.

Next, we discuss the elastic anomaly near T_C. Figure 4 shows the temperature dependence of LA and TA sound velocity with the propagation along the [110] direction. Above T_C, the softening of both LA and TA velocities was observed. For further cooling both velocities markedly decrease. In the typical relaxor SBN, such an elastic anomaly shows broadening near T_C [12]. Figure 5 shows the temperature dependence of sound attenuation coefficient of TA mode with the propagation direction along the [110] direction. The attenuation of TA mode shows the sharp divergence in the vicinity of T_C, and it is the typical behaviour of a nearly second order phase transition.

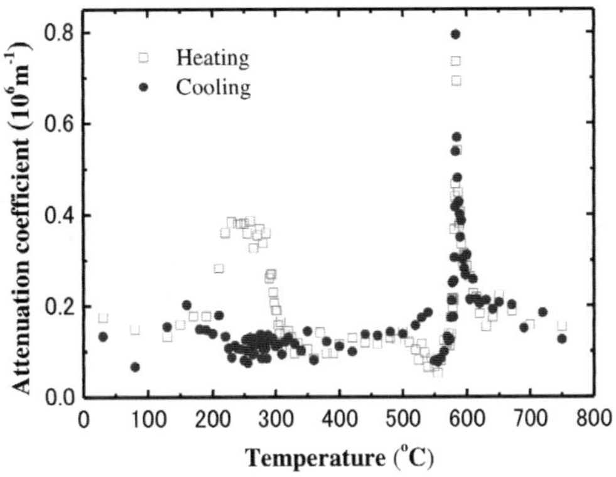

Figure 5. Temperature dependence of the attenuation coefficient of TA mode.

3.3. Ferroelastic incommensurate phase transition

BNN undergoes the incommensurate phase transition at about 300 °C, and spontaneous shear strain appears perpendicular to the *c*-axis [4]. In figure 4 the LA velocity shows the broad minimum near T_{IC} = 285 °C, and the attenuation of TA mode shows a broad peak on the heating near T_{IC} as shown in figure 5. In contrast, on cooling from high temperatures, it is almost constant near T_{IC}. Such a large thermal hysteresis cannot be occurred in a normal ferroelastic phase transition. It is attributed to the typical nature of incommensurate phase transition. The lattice instability of the incommensurate phase occurs near the point [1/4, 1/4, 1/2] at the Brillouin zone, while the soft mode instability occurs at the M-point [1/2, 1/2, 0] which is far from [1/4, 1/4, 1/2]. On heating the incommensurate modulation disappears easily, while on cooling it needs a long time to form the incommensurate modulation. Since the incommensurate structure is connected to macroscopic shear strain in the new type III incommensurate phase, the thermal hysteresis appears strongly in attenuation of TA mode, while not remarkable in that of LA mode.

4. Conclusion

The Brillouin scattering of a multiferroic $Ba_2NaNb_5O_{15}$ crystal was studied to clarify the dynamical properties on both ferroelectric and ferroelastic phase transitions. The LA, TA modes, and central peak were measured in a large temperature range from room temperature up to 750 °C. In the vicinity of a ferroelectric phase transition at about T_C = 585 °C from the prototypic tetragonal *4/mmm* to ferroelectric *4mm* phases, elastic anomaly was observed for LA and TA modes. In addition, the order-disorder nature was confirmed by the observation of critical slowing down. For further cooling elastic anomaly was also observed in the vicinity of a ferroelastic incommensurate phase transition at about T_{IC} = 285 °C into orthorhombic *2mm* phase with the incommensurate modulation. The large thermal hysteresis of elastic anomaly of TA mode near T_{IC} can be attributed the typical feature of the type III incommensurate phase transition predicted recently by Ishibashi and Iwata.

Acknowledgements
One of authors (S. K) is thankful to Prof. Y. Ishibashi on the fruitful discussion about the new type III incommensurate phase in a BNN crystal on the occasion of the 13th International Meeting on Ferroelectricity, Sept. 1-6, 2013, Krakow, Poland.

References

[1] Aizu K 1970 *Phys. Rev.* B **2** 754
[2] Schmid H 1994 *Ferroelectrics* **162** 665
[3] Aizu K 1976 *J. Phys. Soc. Jpn.* **41** 880
[4] Toledano J C, Schneck J and Erradonea G: Ed. Blinc R and Levanyk A P, Incommensurate phases in dielectrics, North-Holland, Amsterdam, 1986. P.233.
[5] Kojima S, Asaumi K, Nakamura T and Minomura S 1978 *J. Phys. Soc. Jpn.* **45** 1433
[6] Kojima S, Nkamura T, Asaumi K, Takashige M and Minomura S 1979 *Solid State Commun.* **29** 779
[7] Ishibashi Y and Iwata M 2013 *J. Phys. Soc. Jpn.* **82** 044703
[8] Yamada T, Iwasaki H and Niizeki N 1970 *J. Appl. Phys.* **41** 4141
[9] Herrero-Albillos J, Marchment P, Salje E K H, Carpenter M A and Scott J F 2009 *Phys. Rev.* B **80** 214112
[10] Kojima S 2010 *Jpn. J. Appl. Phys.* **49** 07HA01
[11] Buixaderas E, Kamba S, Petzelt J, Wada M, Ando S and Tsukamoto T 2000 *Ferroelectrics* **239** 17
[12] Jiang F M and Kojima S 2002 *Phys. Rev.* B **66** 184301

Soliton electric generators based on conducting polymers

Hiromasa Goto

Division of Materials Science, Faculty of Pure and Applied Sciences, University of Tsukuba, Tsukuba, Ibaraki 305-8573, Japan

E-mail: gotoh@ims.tsukuba.ac.jp

Abstract. Iodide doping produces charge carriers in π-conjugated polymers. Motivated by the SSH theoretical model of solitons in one-dimensional conjugated polymers, chemical doping of polyacetylene film is experimentally carried out to generate solitons. An Arago-type wheel electric generator is assembled based on the doped polyacetylene in place of a copper or aluminium plate. This is the first report of electric generation in conducting polymers based on solitons.

1. Introduction

π-Conjugatd polymers consist of alternation of single and double bonds. π-Electrons are delocalized along the main-chains. Polyacetylene is a typical conjugated polymer [1]. It shows metallic refraction because of its highly developed conjugated system, and displays quite high electrical conductivity accompanied by insulator-metal transition by doping [2,3]. Electrical conductivity of the doped polyacetylene is comparable to metals, such as copper. In this case, the charge species of polyacetylene can be characterised as solitons, according to the SSH model proposed by Su, Schrieffer, and Heeger [4].

The spinning-disk-based motor phenomenon was discovered by Arago in 1824. Later, Michael Faraday studied this rotation-induced mechanism [5]. Nikola Tesla investigated this system in his research on alternating current generators and motors [6]. The generator and motor have been constructed with non-ferromagnetic metals such as copper and aluminium. The generators function through movement of free electrons in the metal wheel. The Arago wheel has been employed for induction type energy meter for consumers (for house hold and industrial use). In the present research, the electricity was generated by flow of solitons in the doped polyacetylene (Figure 1). This is the first example of **polymer electric generators** based on charge carriers. The generation of voltage may be related to the magneto-resistance of solitons in the polyacetylene films [7]. Although current from the generator was quite low, electromotive force voltage was achieved.

Figure 1. Polyacetylene

2. Solitons in polyacetylene

Solitons in the polyacetylene can be produced by electron-acceptor doping, as shown in Figure 2. The neutral state of the polyacetylene consists of long extended polyene having no charge carriers. Acceptor vapour-phase doping (iodine) produces polarons in the main-chain. Further doping (heavy doping) allows formation of bipolarons. The polarons and bipolarons behave like solitons in polyacetylene (Figure 2) [8]. In this report, "solutions" for the charge carriers are employed for simplification.

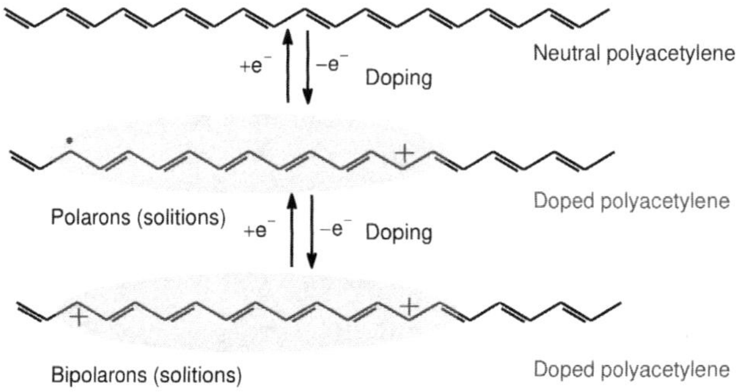

Figure 2. Doping of polyacetylene and generated solitons.

3. Assembly of soliton generator

Based on consideration of polyacetylene as an electrical conductor, assemble of electric generator using polyacetylene was carried out. An Arago disk-type generator was constructed using polyacetylene in place of a metal wheel to demonstrate electric generation by a conducting polymer. Iodine-doped polyacetylene film is attached onto a polyacrylonitrile circular board by adhesive, forming a conducting wheel (diameter: 10 cm). The disk is rotated by an external motor, and the film generates voltage between its centre and edge of the disk, which can be confirmed by an electrometer and an oscilloscope. This instrument demonstrates the possibility that conducting polymers may be used as materials for power generation.

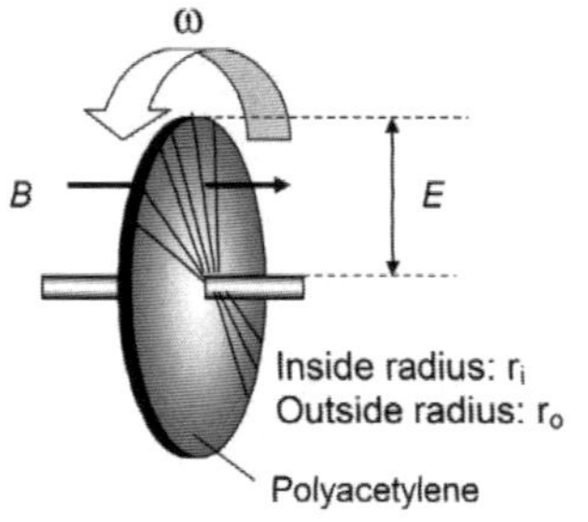

Figure 3. (a) Polyacetylene dynamo system. (b) Polyacetylene disk prepared on a polyacrylonitrile plate (insulator plastic plate).

If a conductive disk rotating at angular frequency ω is subjected to a magnetic flux density *B* (Figure 3), the electromotive force induced in an infinitesimal length Δr can be described as [9]:

$$\Delta E = \mathrm{r}\omega B\Delta \mathrm{r} \quad (1)$$

Integration of the formula from r_i to r_o is

$$\int_{ri}^{ro} E = \omega r B dr = 1/2\omega B(r_o^2 - r_i^2). \quad (2)$$

A fixed-magnetic-type generator was constructed with a neodymium magnet, drive motor, dynamo brush (copper) and thin polyacetylene film on polyacrylonitrile circular-shaped board (insulating plastic plate), as shown in Figure 4.

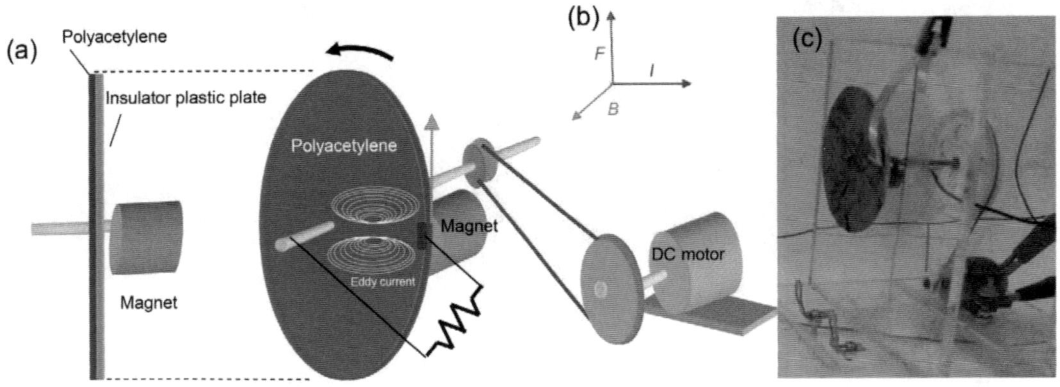

Figure 4. Magnet fixation-type generator based on polyacetylene. (a) Illustration of the entire figure. (b) The Fleming's right hand rule. (c) A photograph of the polyacetylene electric generator.

The magnetic field in the device influences solitons of polyacetylene as the electrically charge particle in the disk, producing electricity perpendicular direction against the magnetic field. The surface contact between the polyacetylene and the dynamo brush in the device is not smooth, resulting unstable generation of electricity. Generated voltage was ca. 0.2 mV (dynamo rotation = ca.1000 rpm measured with a laser revolution indicator). Current from the generator is very low, although the electricity generation function of the device was confirmed. The Fleming's right hand rule can explain direction of current flow (eddy current) across the magnetic field in the device (Figure 4(b)). The Lorentz force from magnetic field can make generation of electricity.

4. Conclusion

An Arago-type polyacetylene dynamo was assembled. Flow of the solitions under the magnetic field affords generation of electricity in the device. To date, electrical conduction of the organic conducting polymers has been well studied. This is the first trial for electric generation in conducting polymers, although practical application of the soliton generator seems to be difficult in the present stage. This study demonstrated that plastics can make electricity.

Acknowledgments

The author would like to express sincere appreciation to Prof. Emeritus Hideki Shirakawa (U. Tsukuba) and alumni of the (former) Shirakawa lab.

References

[1] Ito T and Shirakawa H 1975 *J. Polym. Sci. Polym. Chem.* **13** 1943
[2] Shirakawa H, Louis E J, MacDiarmid A G, Chiang C K and Heeger A J 1977 *Chem. Commun.* 578

[3] Chiang C K, Druy M A, Gau S C, Heeger A J, Louis E J, MacDiarmid A G, Park Y W and Shirakawa H 1978 *J. Am. Chem. Soc.* **100** 1013
[4] Heeger A J, Kivelson S, Schrieffer J R and Su W-P 1988 *Rev. Mod. Phys.* **60** 781
[5] Faraday M, *Faraday's diary (26 December, 1831): being the various philosophical notes of experimental investigation made by Michael Faraday* (London: Bell G and Sons Ltd 1936)
[6] Martin T C, *The inventions, research and writings of Nikola Testla* (CA: Angriff Press, originally published, NY: The Electrical Engineer 1894)
[7] Park Y-W, 2010 *Chem. Soc. Rev.* **39**, 2428
[8] Kudo Y, Kawabata K and Goto H 2013 *J. Phys. Conf. Ser.* **428** 012006
[9] Kato T, Jimbo M 1995 *Bull. Daido. Inst. Tech.* **31** 59

Reversible phase change of new anthracene compounds triggered by the action of light and heat

Hideyuki Kihara and Masaru Yoshida
Nanosystem Research Institute, National Institute of Advanced Industrial Science and Technology (AIST), Tsukuba Central 5, Higashi 1-1-1, Tsukuba, Ibaraki 305-8565, Japan

E-mail: h-kihara@aist.go.jp

Abstract. We synthesized 2-anthroyl and 9-anthroyl ester compounds and investigated their bulk-phase changes associated with photodimerization under melting conditions (melt-photodimerization) and subsequent thermal back reactions. All the five anthroyl compounds exhibited melting points lower than ca. 160 °C, and they were nearly quantitatively converted to the corresponding photodimers by UV irradiation in their molten state. Among them, two 2-anthroyl compounds remained isotropic and lost fluidity during the melt-photodimerization. The obtained photodimers exhibited robust solid-state amorphous phases at room temperature. In contrast, the other three anthroyl compounds showed crystallization during the melt-photodimerization. It was also found that the photodimers returned to the corresponding monomers by heat. We successfully demonstrated rewritable photopatterning using the phase change of the anthracene compounds triggered by the action of light and heat.

1. Introduction

In these days, we treat with a lot amount of digital data at one time; therefore portable data storage media have become more and more important. Among these media, rewritable optical media such as CD-RW and DVD-RW have an advantage in resource saving and reduction of waste. Data recording layer of such media consists of inorganic compounds, such as Ge and Te, known as phase change materials. As long as the inorganic compounds are used as the recording layer, higher-energy-consuming processes, such as, sputtering and CVD are inevitably used for media manufacturing. However, if the recording layer is consisted of organic compounds, more simple processes, such as, spin coating and hot melting are applicable to the fabrication of the media.

It is well known that an anthracene is a typical organic compound that can undergo photodimerization reaction and the obtained photodimer reverts to the original anthracene by light or heat [1,2]. While the anthracene emits fluorescence, the photodimer does not. Using this difference, photopatterning, which can be read by fluorescence observation, was prepared by some researchers [3]. However, there is no example of photopatterning based on phase change associated with photodimerization and decomposition of anthracene compounds. If contrast of a photopatterning is based on phase difference between ordered and disordered phases, the photopattern will have advantage over that based on fluorescence, because the light of any wavelength can be used to read the photopattern based on the phase difference, in contrast, UV light which excites anthracene moiety must be used to read the fluorescence-based photopattern.

In this study, we synthesized five new anthroyl compounds, as shown in figure 1, and investigated their phase behaviour associated with photodimerization and thermal back reaction. It was found that some of the anthroyl compounds having an appropriate substituent showed reversible crystalline and solid-state amorphous phases at room temperature triggered by light and heat. Using this reversible phase change, we successfully demonstrated rewritable photopatterning whose images were visible by cross-polarized observation.

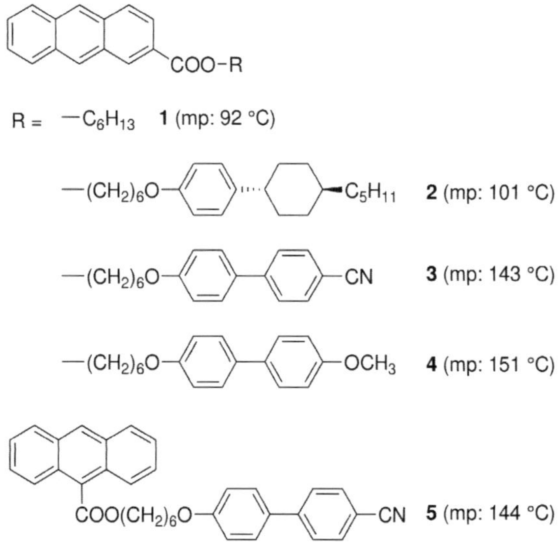

Figure 1. Chemical structures of anthracene derivatives used in this study and their melting points (mp).

2. Experimental

All the anthroyl compounds were synthesized from 2-anthracene carboxylic acid or 9-anthracene carboxylic acid and bromo-terminated alkyl compounds having cyanobiphenyl, methoxybiphenyl, and so on, using 1,8-diazobicyclo[5.4.0]undec-7-ene (DBU) as a catalyst for esterification [4]. ^{1}H NMR spectra were recorded on a JEOL JNM-ECX400. The thermal properties of the samples were evaluated by polarized optical microscopy (POM) using an Olympus BH2 equipped with a Mettler FP82HT hot stage and by differential scanning calorimetry (DSC) using a Seiko Instruments Inc. DSC 6200. Heating and cooling rates were 10 °C min^{-1}. Photoirradiation of the samples was carried out using monochromatic 365-nm UV light (5 mW cm^{-2}) from a 500-W high-pressure mercury lamp through optical filters. During the photoirradiation, the temperature of the samples was controlled by a Likam LK 600PH hot stage. Gel permeation chromatography (GPC) was conducted using a Shimadzu system (an LC-10ADvp pump unit, an SPD-10Avp UV detector (detection wavelength: 280 nm), a CTO-10Avp column oven and an SCL-10Avp controller) with Shodex KF800D and KF805L columns, using THF as the eluent. High-performance liquid chromatography (HPLC) was performed by a TOSOH system (a CCPM-II pump unit, a PD-8020 UV detector (detection wavelength: 290 nm), a CO-8020 column oven, and a PX-8020 controller) using a TSK silica-60 column. The mixed solution (toluene/dichloromethane/2-propanol = 90:10:0.2 v/v/v) was used as the eluent for the HPLC experiments.

3. Results and discussion

Non-substituted anthracene and its photodimer have symmetric structure and thus have significantly high melting points at 216 [5] and 260 °C (accompanied by decomposition) [6], respectively. Such high melting points prevent us from applying the non-substituted anthracene to phase change materials. Therefore, we synthesized new anthracene derivatives. As shown in figure 1, the melting points of the anthracene compounds have been effectively lowered by substitution.

Photoirradiation of the anthracene derivatives were carried out using Linkam hot stage and high pressure mercury lamp. 2 mg of the anthracene compounds were placed between glass plates and irradiated with UV light (365 nm, 5 mW cm^{-2}). If the molten sample of anthracene compound, **3** or **4**, was cooled to room temperature without photoirradiation, it naturally returned to crystalline phase. However, when the sample was irradiated with UV light for more than 5 min in the molten state, it did not return to crystalline phase even if cooled to room temperature, instead solid-state amorphous phase was formed. This solid-state amorphous phase could be prepared by not only rapid cooling but also slow cooling after UV irradiation. This was the indication that the obtained amorphous phase was thermodynamically stable state. In contrast, UV irradiation in the molten state of the remaining anthracene compounds, **1**, **2**, and **5**, resulted in the crystalline phase formation.

As above-mentioned, there were two kinds of phase behaviour in photoirradiation of the anthracene compounds in the molten state. One was solid-state amorphous phase formation and the other was crystalline phase formation. In both cases, we confirmed the dimer formation during UV irradiation in the molten state by means of GPC measurements. As the UV irradiation time increased, the peak associated with photodimers increased, instead the peak of anthracene monomers decreased. We found that the anthracene monomers converted to photodimers almost quantitatively in 20 min of UV irradiation. As a representative, GPC result obtained by melt-photoirradiation of **3** is shown in figure 2.

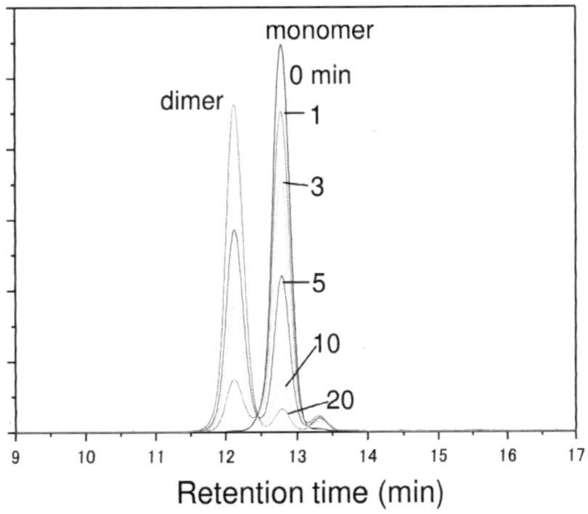

Figure 2. GPC charts of **3** obtained by UV irradiation in the molten state at 150 °C for various periods of time.

In addition, we found that the back reaction, namely, decomposition from dimer to monomer occurred by heating of the anthracene photodimers. We investigated the effect of heating temperature on the thermal back reaction by GPC analysis and found that the higher the heating temperature became, the sooner the back reaction from dimer to monomer proceeded. When the heating temperature exceeded to 200 °C, the photodimers reverted to the corresponding monomers almost quantitatively in 5 min. Interestingly, the reproduced monomers from amorphous-phased photodimers of **3** and **4** showed crystalline phase again at room temperature. NMR measurements also revealed the photodimerization in the molten state of the anthracene compounds and the subsequent thermal back reaction almost quantitatively occurred. Table 1 summarizes the phase change behaviour of the anthracene compounds associated with photodimerization and thermal back reaction. The relationship between the structure of anthracene compounds and phase change behaviour have not been clarified yet, because the data obtained so far are too limited to explain.

Table 1. Phase state of the photodimers[a] and the reproduced monomers[b] at room temperature.

Anthracene derivatives	Phase of the photodimer at room temp.	Phase of the reproduced monomer at room temp.
1	Crystalline	Crystalline
2	Crystalline	Crystalline
3	Solid-state amorphous	Crystalline
4	Solid-state amorphous	Crystalline
5	Crystalline	Crystalline

[a] Obtained by UV irradiation in the molten state.
[b] Obtained by thermal back reaction.

Note that all the photodimers in table 1 were obtained by UV irradiation in the molten state. In fact, while UV irradiation of anthracene compound **3** in the crystalline phase also gave the photodimer, the phase of the obtained photodimer was crystalline phase. In order to investigate the difference between the two photodimerization conditions, we next carried out HPLC measurements. There is a possibility that asymmetric anthracene compound produces regio-isomers of photodimer in the dimerization reaction as shown in figure 3[7]. Unlike GPC analysis, HPLC measurement can detect the isomer formation under appropriate measuring condition. As shown in HPLC chart obtained by the UV irradiated sample of **3** (figure 4a), only one isomer was formed in the UV irradiation in the crystalline state, however, at least three isomers were produced in the UV irradiation in the molten state (figure 4b). From this result, it could be speculated that the presence of regio-isomers was a necessary condition for the formation of the solid-state amorphous phase for the photodimers. In addition, we found that the presence of regio-isomers is one of the necessary conditions but not sufficient condition from the next experimental results. HPLC analysis was also performed for the photodimerization reaction in the molten state of compounds **1** and **2**. Although they formed crystalline phase during UV irradiation as already mentioned, we found that the obtained photodimers were mixtures of regio-isomers. This result indicated that the presence of the regio-isomers in the photodimer was necessary condition but not sufficient one for the formation of solid-state amorphous phase.

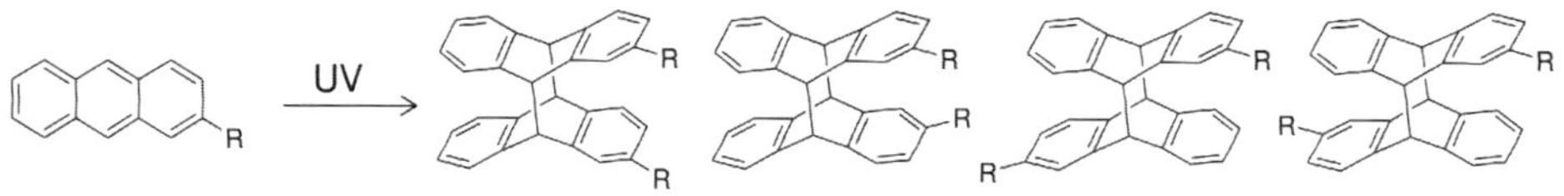

Figure 3. Chemical structures of possible photodimer region-isomers obtained from 2-anthroyl compounds.

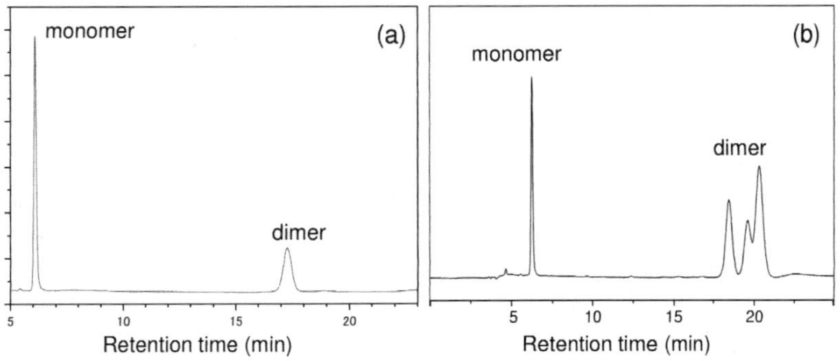

Figure 4. HPLC charts of **3** measured after UV irradiation for 10 min in the crystalline phase (a) and in the molten phase (b).

In this stage, we think the fluidity of the photodimer is another one of the necessary conditions for the formation of the solid-state amorphous phase. A schematic illustration in figure 5 shows our speculation. When anthracene monomer was irradiated with UV light in the molten state, the obtained photodimer showed amorphous phase for a while after its formation. And then if the photodimer has enough fluidity at the irradiation temperature, it will organize into crystalline phase because the melting point of the photodimer is higher than the irradiation temperature. In fact, we confirmed the crystalline photodimers had the melting points at temperature higher than 200 °C. In contrast, if the photodimer does not have enough fluidity at the irradiation temperature, it cannot change to ordered phase and the amorphous phase will be immobilized. In order to clarify the mechanism of the solid-state amorphous phase formation, we are now doing computer simulation.

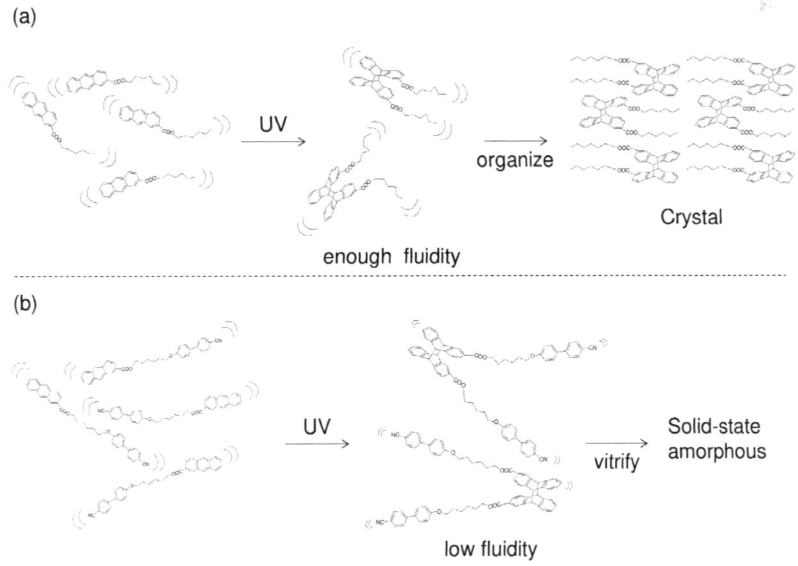

Figure 5. Schematic illustration for plausible mechanism of crystalline (a) or solid-state amorphous (b) phase formation in melt-photodimerization of the anthracene compounds.

As mentioned so far, we found that the some anthracene compounds can be converted to the photodimer which shows solid-state amorphous phase by light and heat, moreover, the photodimer can be returned to the crystalline monomer by heat. Therefore, next we tried to realize rewritable photopatterning based on the phase contrast using our anthracene compound **3**. In order to record a photopattern, the compound **3** was melted to amorphous phase and irradiated with UV light through the photomask, and then the sample was cooled to room temperature. When the obtained sample was observed through crossed polarized light, we could see a pattern faithfully reflecting the photomask (figure 6a). With the careful observation of the photopattern, it was found that the dark part corresponded to the UV-irradiated part, and the bight part to the unirradiated part. Generally, in the cross polarized light observation, the light can penetrate through the ordered-phased region but not the amorphous-phased region. Therefore, the fact that the UV-irradiated parts composed the dark areas of the photopattern was consistent with the phase change from crystalline to solid-state amorphous phase. The photopattern can also be seen by fluorescence observation (figure 7). Compared to the two observations, cross polarized light observation is superior because the light of any wavelength can be used, in contrast, UV light which can excite anthracene moiety should be used in the fluorescence observation and such UV light gradually damages the photopattern.

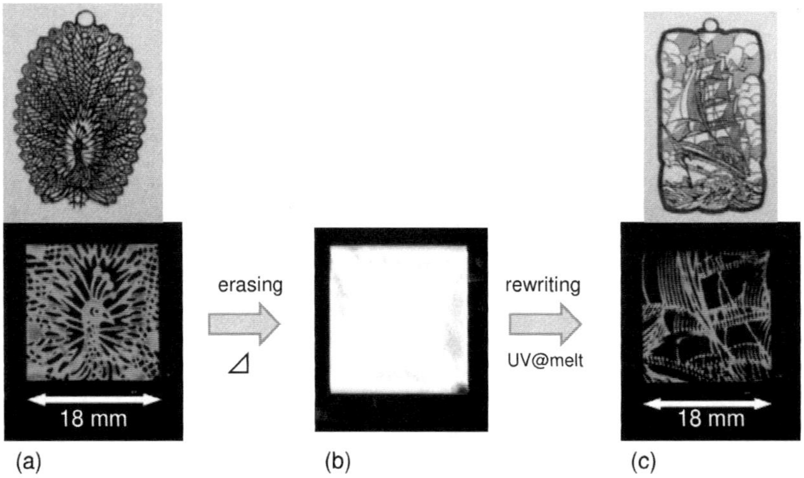

Figure 6. (a) Cross-polarized-light image written in **3** and a photomask used for photopatterning. (b) Cross-polarized-light image of the erased sample. (c) Cross-polarized image written in the reproduced sample and a photomask used for rewriting another photopattern.

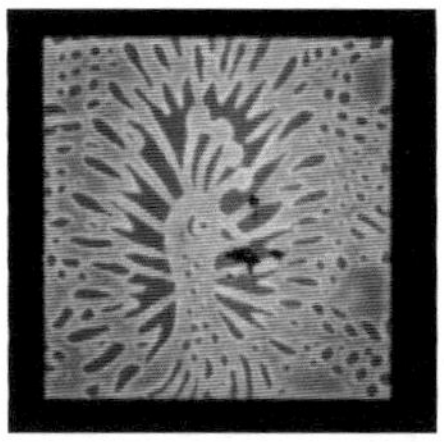

Figure 7. Fluorescence image of the same sample in figure 6a.

We also found that it could be possible to erase the photopattern by heating the sample at about 200 °C. The photodimer in the dark areas revered to the anthracene monomer by thermal back reaction, and the obtained monomer showed crystalline phase, consequently the whole sample became bright in cross polarized light observation (figure 6b). In addition, it was confirmed that rewriting the photopattern in the erased sample was possible (figure 6c). Unfortunately, this rewriting process can be repeated in several times because thermal decomposition gradually occurred.

4. Conclusion

We synthesized new anthracene compounds and found that if the anthracene had an appropriate substituent, its photodimer obtained by UV-irradiation in the molten state showed solid-state amorphous phase. The obtained solid-state amorphous phase was very robust at room temperature, however, the photodimer reverted to the original monomer when heated at about 200 °C and the reproduced monomer showed crystalline phase. Using the reversible phase change of the anthracene compounds, we successfully demonstrated rewritable photopatterning by the action of light and heat.

Acknowledgments

This work was partially supported by a KAKENHI (235560221) Grant-in-Aid for Scientific Research (C).

References

[1] Bouas-Laurent H, Castellan A, Desvergne J P and Lapouyade R 2000 *Chem. Soc. Rev.* **29** 43
[2] Bouas-Laurent H, Castellan A, Desvergne J P and Lapouyade R 2001 *Chem. Soc. Rev.* **30** 248
[3] Rameshbabu K, Kim Y, Kwon T. Yoo J and Kim E 2007 *Tetrahedron Lett.* **48** 4755
[4] Blatch A E, Fletcher I D and Luchhurst G R 1995 *Liq. Cryst.* **18** 801
[5] Karl N, Heym H and Stezowski J 1985 *Mol. Cryst. Liq. Cryst.* **131** 163
[6] Komatsu K, Tsuji K and Takeuchi K 1989 *Tetrahedron Lett.* **30** 4689
[7] Ishida Y, Kai Y, Kato S, Misawa A, Amano S, Matsuoka Y and Saigo K 2008 *Angew. Chem. Int.Ed.* **47** 8241

What conductive polymers have taught us about the meaning of education: education before innovation

Kazuaki Hiroki

Tsuyama National College of Technology

Numa 624-1, Tsuyama, Okayama 708-8509, Japan

E-mail: hiroki@tsuyama-ct.ac.jp

Abstract. There are many scientists and engineers who made a great discovery or a breakthrough invention. What is common to them? It is that they had a chance to experience a deep impression of science and technology in their childhood. Science shows and experiment classes are extremely effective methods to help students learn the excitement of Chemistry. In this paper, I discuss the importance of education in chemistry by presenting an illustration about the development and implementation of experiment classes on conductive polymers.

1. Introduction

Several years ago, I produced exhibitions and experiment labs as a member of National Museum of Emerging Science and Innovation in Tokyo (the Miraikan) [1]. As soon as I arrived to the Miraikan, I joined the renovation team for the exhibitions. There were exhibitions about nanotechnology originally, which had a problem of becoming obsolete quickly because of the fast change in this area. So our team decided to make an exhibit about how breakthrough inventions, discovery and technological innovations happen. We named the exhibitions "The driving forces of innovations". Through the research for this exhibition, I met many scientists and engineers, and I wondered: What is a driving force of innovation that brings a revolutionary inventions and discoveries? What is common to the Japanese Nobel Prize in Chemistry winners since 2000? I found the answers to these questions. What is common to the great scientists is that they became familiar with nature, and had impressive experiences with nature in their childhood. They were impressed to watch flowers in full glory, trees of beautiful green, stars glittering, blue of the sky, the depth of the sea, red fiery autumn leaves, and so on [2]. Or they were strongly impressed by exciting scientific experiments. Some were excited by chemical reactions, which caused sudden color changes, emissions, or sometimes explosions. It goes without saying that they never forgot the excitement, and went on to learn chemistry, became scientists or engineers, and then engaged in creative research and innovation. In some instances, the importance of Serendipity is acknowledged, such as in the invention of conductive polymers by Prof. Hideki Shirakawa [3].

Therefore "impression" is one the keywords. So, what is an impressive method to tell people about the wonderfulness of chemistry? An answer is a science show or experiment class. These are extremely effective methods to help students experience the excitement of chemistry. Below, as examples of good experiment classes, I present two experiment classes on conductive polymers.

2. Result and discussion

2.1. Experiment class in Conductive polymers for general people

Prof. Hideki Shirakawa was awarded the Nobel Prize in Chemistry for the discovery of conductive polymers in 2000. But even before that auspicious year, he was carrying the fear that the opportunity for chemical experiments in schools was decreasing. Unfortunately, this seemed to cause a "hate science" attitude among students in Japan. So Prof. Shirakawa and the Miraikan planned and carried out the science workshop on the theme of conductive polymers at the pace of almost one a month [1]. Because it is the "mirai"="future" science museum, we have to exhibit advanced science, and the experimental classes were not an exception.

In the workshops, Prof. Shirakawa directly teaches the participants. The target audience are all fifth-grade elementary school students, as well as higher grade students and adults. As a Shirakawa-Kijima laboratory graduate and also as a science communicator for the Miraikan, I was responsible for the development and implementation of the experimental classes. And I had noticed that conductive polymers have immeasurable possibilities as a teaching material [4].

Experiments with conductive polymers may seem to be too difficult for general people. However, conductive polymers actually possess various characteristics that can make chemical experimental workshops very attractive. These characteristics are:

1. They can be relatively easily synthesized.
2. Doping conditions induce drastic changes in their colors and characteristics.
3. Doping induces various physical properties, such as electrical conductivity, electrochromism, and electroluminescence.
4. Their characteristics are all linked directly to various applications.

In addition, experiments using conductive polymers allow people to experience the synthesis and application of the polymers at the same time. They are rare materials that can contribute to learning the wonder of the entire spectrum of science, in addition to chemistry. Below, I demonstrate the use of conductive polymers for scientific education.

2.2. Experiment Classes in Conductive Polymers(1)

Using PEDOT to fabricate a transparent film speaker. Poly(3,4-ethylenedioxythiophene) (PEDOT) is an important conductive polymer, which has attracted attention for its unique properties [4]. Its excellent doping performance, high stability and conductivity make PEDOT useful for a variety of applications, including organic electroluminescence devices (OLEDs), photovoltaic cells, sensors, etc [5] [6]. One of the most interesting applications is a transparent film speaker, which has already been commercialized by the Korean company “Fils” [7].

Prof. Shirakawa and the Miraikan Science Work Shop (Miraikan SWS) planned and performed the experiment classes using PEDOT transparent film speakers. In this section, I explain the method and contents of these classes.

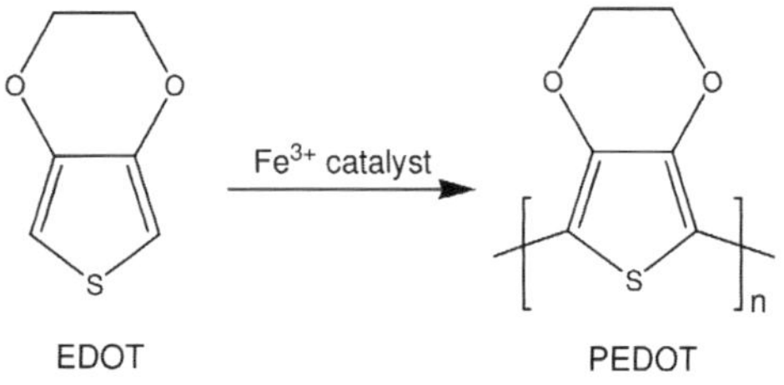

Scheme 1. Oxidative polymerization of EDOT using catalyst.

3,4-Ethylenedioxythiophene (EDOT) is polymerized on the both sides of piezoelectric film using Fe^{3+} catalyst (Scheme 1). We usually use Fe^{3+} catalyst, such as an ethanol solution of iron (III) *p*-toluenesulfonate (Figure 1). Extra mononer and catalyst are removed by rinsing with ethanol (Figure 2). Thin PEDOT electrodes are re-doped with the polymer dopant poly (vinylsulfonic acid), or PVS, by applying a solution of PVS (Figure 3). When we attach copper foils as auxiliary terminals, the transparent film speaker is completed, with its thin and mysterious blue colored PEDOT/PVS electrodes (Figure 4). When the electrodes are connected to an audio player (Figure 5), the flexible piezoelectric film vibrates according to the audio signals (Figure 6).

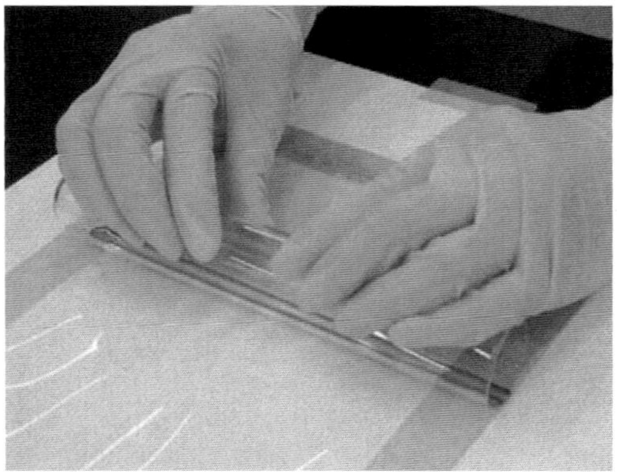

Figure 1. The steps to form a thin PEDOT electrode on a piezoelectric film by applying a solution of EDOT and catalyst.

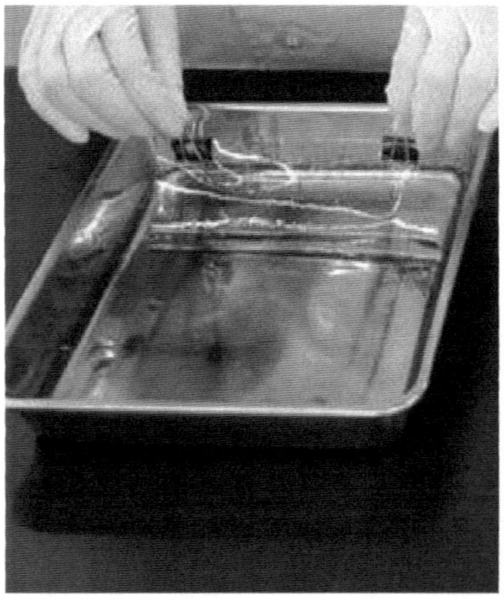

Figure 2. The steps to removed extra mononer and catalyst by rinsing with ethanol.

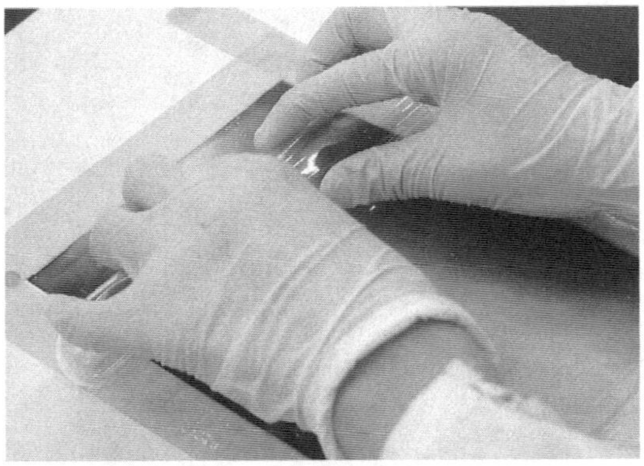

Figure 3. The steps to re-dope thin PEDOT electrodes with the polymer dopant PVS, by applying a solution of PVS.

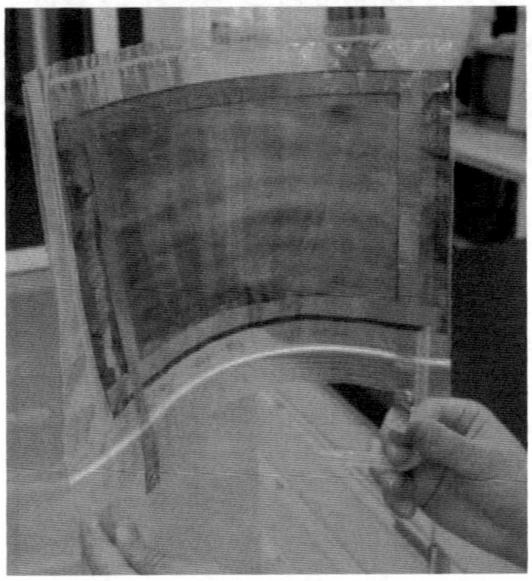

Figure 4. The steps to attach copper foils as auxiliary terminals, and complete the transparent film speaker.

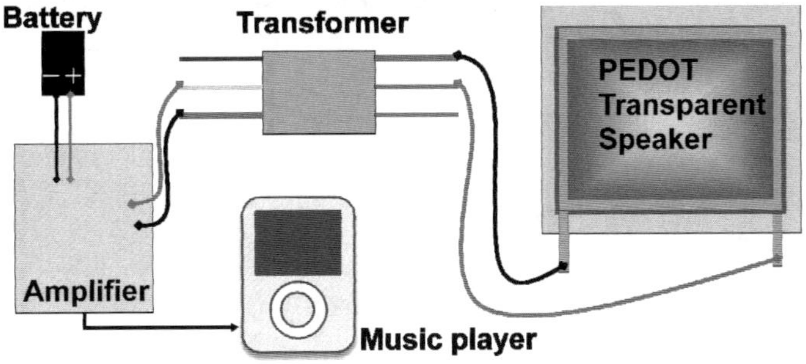

Figure 5. The steps to connect copper foil terminals to an audio player.

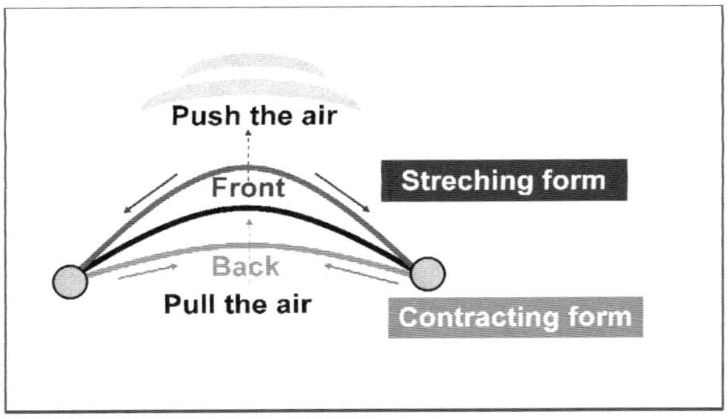

Figure 6. Mechanism of the flexible transparent film speaker.

This means that PEDOT plays the role of a transparent thin film electrode, converting the vibration audio voltage, which is the trick to make the sounds come out. When lovely music flows from the film speakers, everyone will smile.

2.3. Experiment Classes in Conductive Polymers(2)

A HANDMADE Polymeric Organic Electroluminescent Device [OLED]. As seen in the experiment above, PEDOT is a useful conductive polymer. It can be considered one of the ideal molecular structures of conductive polymers. Another important conductive polymer is Poly [2-methoxy-5-(2-ethylhexyloxy)-1,4-phenylenevinylene](MEH-PPV), which applied for a light-emitting layer of polymeric OLEDs.

Prof. Shirakawa and the Miraikan SWS also planned the experiment "fabrication of HANDMADE polymeric OLEDs." In this section, I explain the method and contents of the experiment classes of fabricating a polymeric OLED using PEDOT and MEH-PPV (Figure 7).

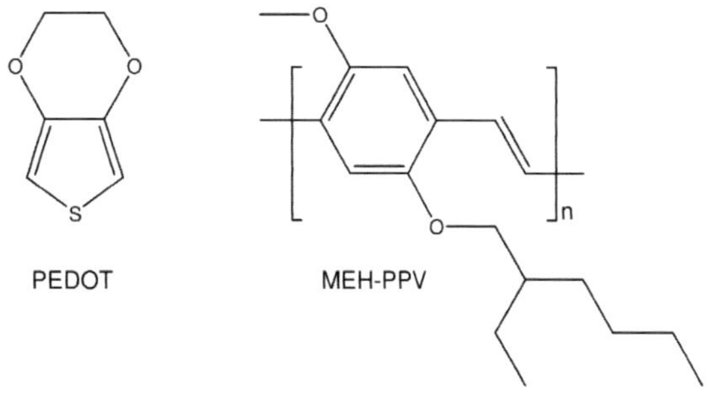

Figure 7. PEDOT (a hole injection / transport layer) and MEH-PPV (a light-emitting layer).

What do I mean by "HANDMADE"? The use of evaporator and spin coater is difficult for children, and in the first place takes an unreasonable amount of time.

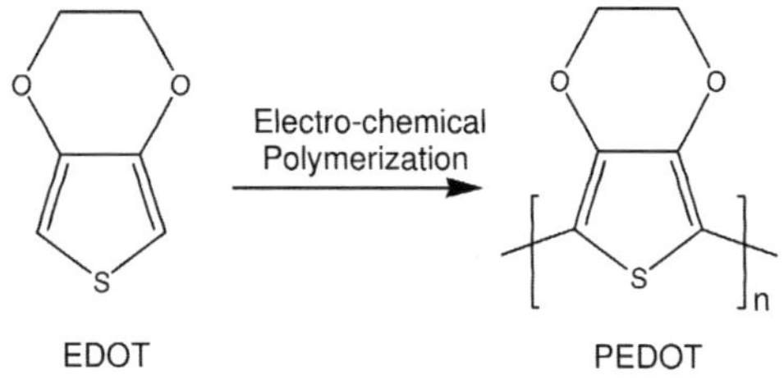

Scheme 2. Electro-chemical polymerization of EDOT.

So we make PEDOT layer on top of the ITO glass by the electro-chemical polymerization of EDOT (Scheme 2, Figure 8), and dip coating the toluene solution of MEH-PPV layer on it (Figure 9). Next, the solvent is removed by drying with a hair dryer. On another ITO glass, we place the Ga-In eutectic (Figure 10). Finally we stick them together with double-sided tape with a hole (Figure 11), and the HANDMADE polymer organic EL device is completed (Figure 12) [8].

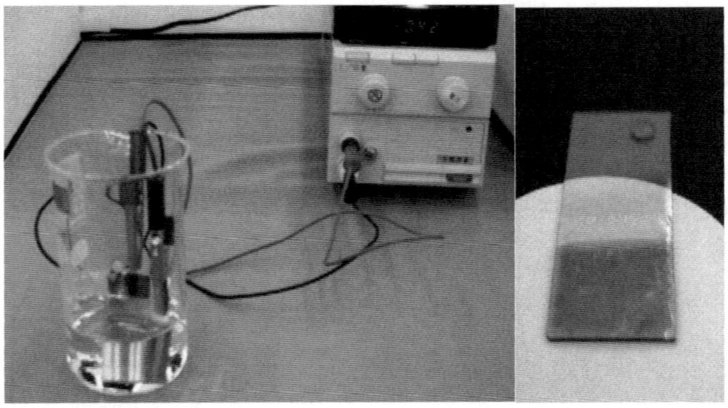

Figure 8. The steps to synthesize a PEDOT as the hole injection layer on an ITO glass by electro-chemical polymerization.

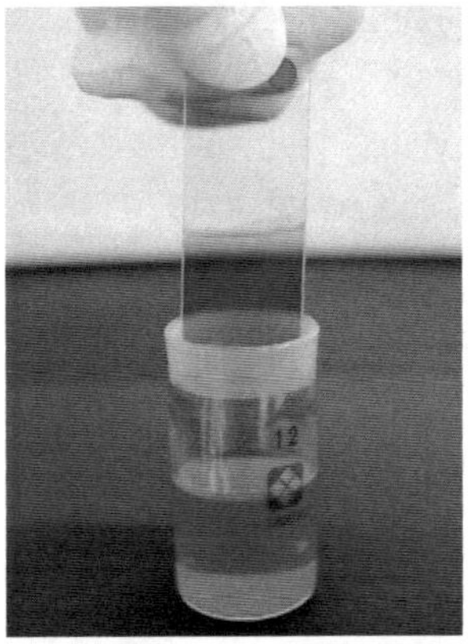

Figure 9. The steps to dip coat the toluene solution of MEH-PPV on the ITO glass with thin PEDOT layer.

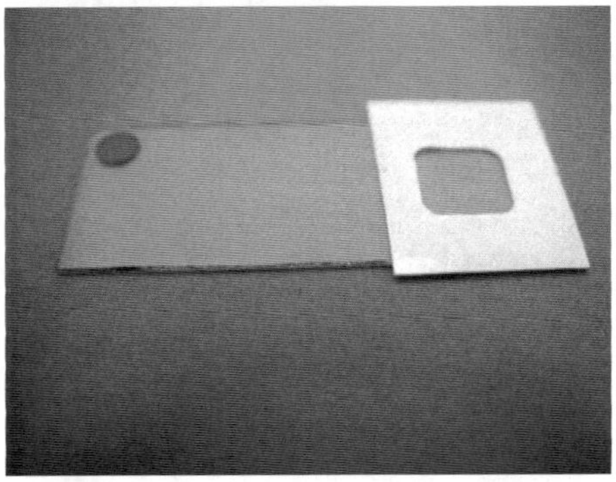

Figure 10. The steps to put the Ga-In eutectic as the electron injection layer on another ITO glass.

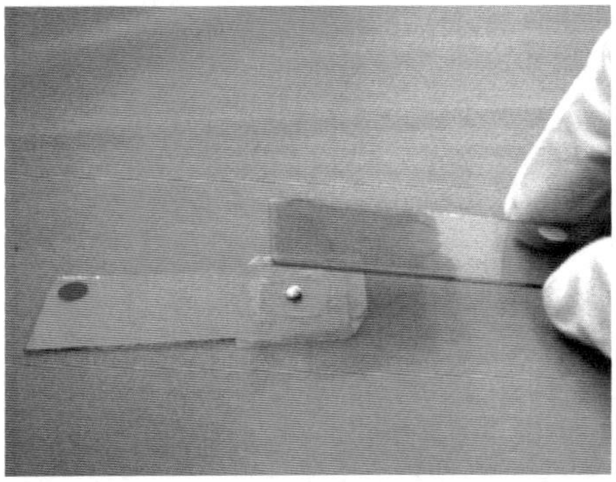

Figure 11. The steps to stick together two ITO glasses with double-side tape.

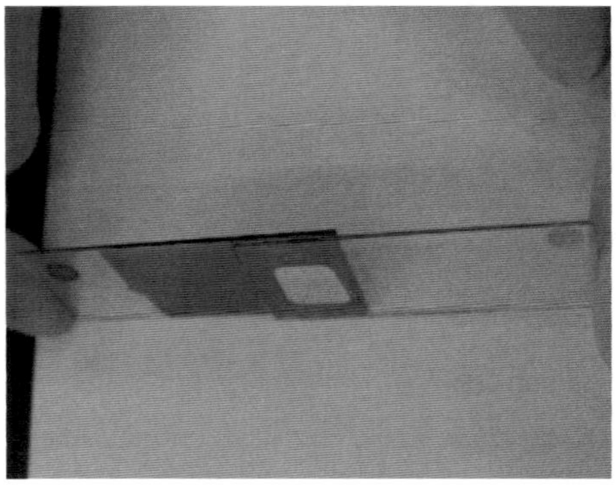

Figure 12. Completed "HANDMADE" polymeric OLED.

The construction of this HANDMADE polymeric OLED is described in Figure 13. It is a typical multi-layer structure, ITO (electrodes), PEDOT (a hole injection / transport layer) , MEH-PPV (a light-emitting layer), and Ga-In eutectic (an electron injection / transport layer).

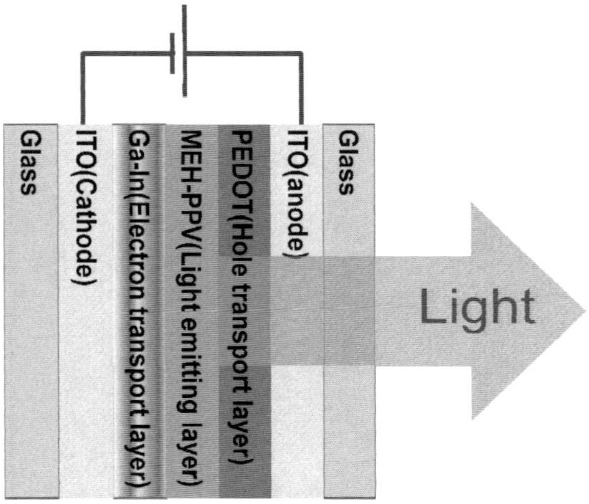

Figure 13. Construction of a HANDMADE polymeric OLED.

Thus, we developed methods to fabricate a HANDMADE polymer OLED without either a spin coater or a vacuum deposition machine. When the electrodes are connected to a power supply, the handmade polymeric OLED emits light orange (Figure 14). When the OLED lights up, with its warm orange light, everyone is sure to be impressed.

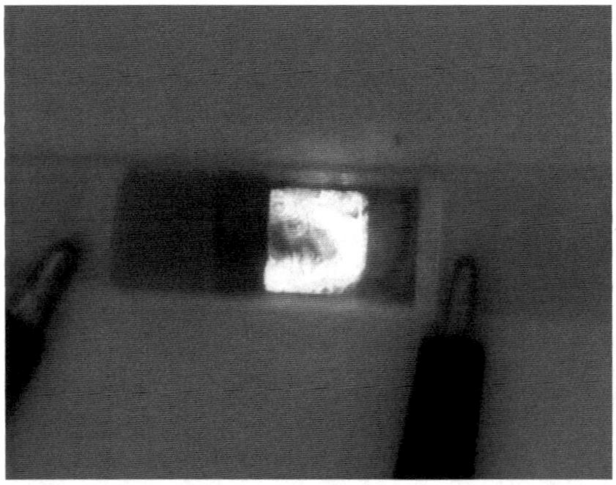

Figure 14. Light emitting the HANDMADE polymeric OLED, in the experiment class at Nagoya University (2012).

3. Summary

We have developed interesting and exciting experiment classes on conductive polymers. The experiments to fabricate devices are simplified as much as possible; so that the participants can succeed easily, especially the children. From one point of view, these experiments can be regarded as simplified industrial processes of making devices, and may help students understand any featured processes, such as roll-to-roll processing, printable electronics, etc.

On the other hand, from educational point of view, we have succeeded in developing attractive experiment classes on advanced materials.When I see the smiles and excitement of the participants, I am always happy. The purpose of the experiment classes is to give knowledge and carry out the experiment itself, but both of these are secondary. The true purpose of the experiment is demonstrating the excitement of conductive polymers. Even if students fail the experiment, it is still exciting. This can be the magic that inspires students. The experience of being excited by chemistry provides a chance to aspire greater heights in science and engineering. It is no wonder that the scientists and engineers who stir up the innovation for the future begin with curiosity and wonder.

The true meaning of education is having the students watch, listen, be surprised, and experience the pleasure of understanding, rather than giving the knowledge unilaterally. We will continue to develop a new experiment classes on conductive polymers in the future. We will continue to verify the effectiveness of both the scientific process and the educational effect.

Acknowledgments

The author wishes to acknowledge the assistance and encouragement from Prof. Hideki Shirakawa (Professor Emeritus of University of Tsukuba) and science communicators of the Miraikan (National Museum of Emerging Science and Innovation, Japan).

References

[1] http://www.miraikan.jst.go.jp/en/ "Special Course: Let's synthesize Conductive Plastics"

[2] http://www.kahaku.go.jp/english/ "Messages from Nobel Prize in Chemistry winners"

[3] Shirakawa H 1990 Modern Alchemy: Plastics that conduct electricity *ILLUME* Number 4 Vol 2 No 2 (Tokyo:Tokyo Electric Power co. Inc.) 22

[4] Hiroki K 2010 Lecture on Experiments of Conductive Polymers *Chemistry & Chemical Industry* **63** 808

[5] Elschner A, Kirchmeyer S, Lovenich W, Merker U and Reuter K 2010 *PEDOT: Principles and Applications of an Intrinsically Conductive Polymer* (Boca Raton, FL : CRC Press)

[6] Hiroki K and Shirakawa H, 2012 PEDOT application for Chemistry Education *Material Properties of PEDOT and Its Application to Devices*, ed Okuzaki H (Tokyo: Science & Technology) chapter 11 pp 399–405

[7] http://fils.co.kr/

[8] Shirakawa H and Hiroki K 2012: *The Fundamentals of Conductive Polymers (Material Matters Basics* vol 6*)* (Tokyo: Sigma-Aldrich Japan)

IOP Conf. Series: Materials Science and Engineering **54** (2014) 012022 doi:10.1088/1757-899X/54/1/012022

The international workshop on science and patents, its origin and purpose

Rafaël H.L. Kiebooms[1] and Hiromasa Goto[2]

[1] European Patent Office, Patentlaan 2-9, 2288EE The Hague, The Netherlands
[2] Division of Materials Science, Faculty of Pure and Applied Sciences, University of Tsukuba, Tsukuba, Ibaraki 305-8573, Japan

rkiebooms@epo.org, gotoh@ims.tsukuba.ac.jp

Abstract. Since 2009 lectures on intellectual property in an international context were developed for an audience of researchers and engineers with a focus on patents and positioning the process of acquisition in an international context. The didactical approach puts a lot of value on the application of the principles of intellectual property acquisition to the actual personal research topics of the students. In parallel with the development of the intellectual property lectures, the International Workshop on Science and Patents (IWP) was set up to provide a discussion forum for researchers and engineers. The purpose is to give the inventors themselves an opportunity to present their experience with the process of applying for patents.

1. Introduction

In 2009 at the occasion of the bilateral collaboration project [1] between the European Patent Office (EPO) and the Japanese Patent Office (JPO) Prof. Dr. Hiromasa Goto expressed the need to provide lectures on intellectual property at the Tsukuba University. These lectures should focus on the following aspects: be addressed at scientists and engineers, concentrate on patents and position the whole in an international context. In addition he was wondering whether it would be possible to include in such lectures practical group work using on-going research topics to provide a hands-on environment for improved understanding of various aspects of intellectual property (IP) protection.

2. The International Workshop on Science and Patents (IWP)

Indeed, few lectures address the lack of knowledge of the inventors themselves on the process of acquiring intellectual property protection in an international context. Hence, since 2009 we developed a series of lectures with the goal of providing scientists and engineers with the essential knowledge about their role in the process of protecting their intellectual property at the global level. These lectures are since 2013 officially part of the PhD programme of the Institute of Materials Science at the Tsukuba University.

In parallel we realised that there is a need to offer a forum for researchers to present their experience with the process of obtaining patents and discuss how the researchers experienced and addressed the various problems, which emerged during this process. In contrast to the usual conferences and meetings for patent attorneys, lawyers, economists, policy makers and managers, the IWP discussion forum therefore aims at giving a voice to scientists and engineers, who are at the core of any invention.

The earliest interaction with the Tsukuba University goes back to 1996 at the International Conference on Science and Technology of Synthetic Metals (ICSM), held at Snowbird, (Utah, US). At that time I was carrying out post-doctoral research as a researcher from the National Science Foundation - Flanders at the Institute for Polymers and Organic Solids at the University of California Santa Barbara, renamed the Center for Polymers and Organic Solids (CPOS). The center, founded in 1982 as Institute for Polymers and Organic Solids (IPOS) by Prof. Dr. Alan Heeger and Prof. Dr. Fred Wudl, was converted to CPOS in 2000. CPOS combines interdisciplinary research in the fields of physics, chemistry, polymer science and biology. Expert knowledge from these disciplines provides the basis for a fundamental research programme aiming at developing new materials with functional properties. These materials are not only small molecules, but also conjugated organic polymers characterized by electrical conductivity, anisotropic linear and nonlinear optical properties, and novel electrochemical properties [2].

During my research stay at IPOS in 1996, I was strongly interested to expand my earlier research on polyisothianaphthene to other isothianaphthene based low band gap polymers. At ICSM 1996 Prof. Dr. Goto reported on his research combining polythiophene methines and polypyrrole methines with liquid crystal side chains [3]-[4]. This work built on the first results on polythiophene methines as reported by Chen and Jenekhe [5]-[6]. After sharing our experience on the synthesis of low band gap polymers, I returned to the lab and prepared the first polyisothianaphthene methine [7].

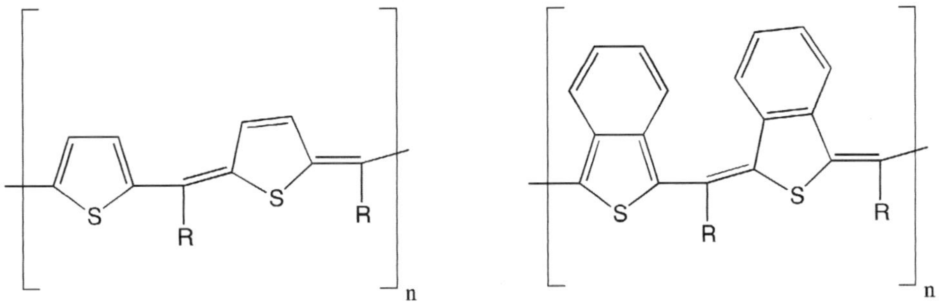

Figure 1. Polythiophene methine and polyisothianaphthene methine

Following the successful synthesis of a soluble and stable polyisothianaphthene methine the very first collaboration with the group at the Tsukuba University was set up. The group at the Tsukuba University was at that time still headed by Prof. Emeritus Dr. Hideki Shirakawa. In 1999, about one year before Prof. Emeritus Dr. Hideki Shirakawa was awarded the Nobel price 2000, I went to the Tsukuba University with a grant from the National Science Foundation - Flanders and a foreign professor grant from the Tsukuba University. The combination of my knowledge on isothianapthene chemistry and the expertise on liquid crystals developped by Prof. Dr. Goto and Prof. Dr. Kazuo Akagi resulted in the synthesis of a series of polyisothianaphthene methines with various liquid crystalline side groups [8].

At the level of research our collaboration has continued with the synthesis of other low-bandgap polyisothianaphthene methines such as chiral polyisothianaphthene methines having magnetic properties [9] and containing a chiroptically active substituent [10]. In the early phases of the development of the intellectual property lectures, the research topic on conducting polymers was also used as an excellent opportunity to discuss the practical aspects of intellectual property protection.

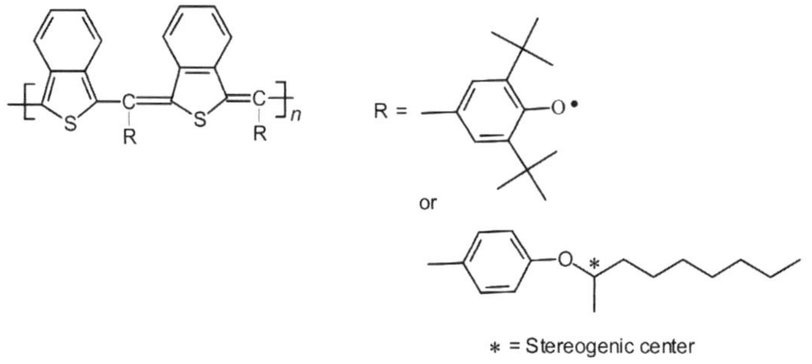

Figure 2. Polyisothianaphthene methines having a stable radical group or a chiroptically active substituent.

In 2001 I took up a position at the EPO to search and examine patent applications among others in the field of conducting polymers. During the years following my start at the EPO I stayed in touch with the group at the Tsukuba University. Over time we came to realize that there is a strong need among scientists for more detailed knowledge about intellectual property, at the national and the international level, and the role played by scientists and engineers. The bilateral collaboration project [1] between the EPO and the JPO provided a fertile ground to develop together with Prof. Dr. Goto new ideas and approaches to teach intellectual property in an international context.

3. Scientists and Engineers

The need to address an audience of scientists and engineers as expressed by Prof. Dr. Goto finds its origin in that most lectures on intellectual property and patents are aimed at an audience of intellectual property representatives, such as patent attorneys, people with a back ground in law and managers of companies.

However, scientists and engineers play an often underestimated key role during the prosecution of the patent application. First of all they are the actual creators of the invention. Secondly, they will provide the technical contents for the patent document. Finally they will provide scientific support to the patent attorney during substantive examination of the patent application.

It is therefore important that scientists and engineers know at which level they contribute in the overall management of an intellectual property portfolio. Furthermore, scientists and engineers also need to understand how the criteria for patentability are assessed by a patent office. And finally, they need to be aware of the various stages of the patent application procedure.

4. Protecting Intellectual Property by means of patents

Successful business needs IP protection based on a combination of IP rights (e.g. copyrights, trademarks, design rights and patents), which provide detailed definitions of technical and artistic ideas and concepts. This IP rights portfolio is used as a basis for value assessment of the intellectual assets of a company [11].

There are many reasons why the application for intellectual property rights may not be successful. Even when IP rights have been granted, the utilization of these rights may not provide the anticipated results. An appropriate IP strategy should consider the position of the organization in its socio-economic ecosystem, the various possibilities to acquire intellectual property rights and finally how to benefit from the utilization of the so-acquired intellectual property rights. Establishing such a holistic strategy requires the close collaboration of experts not only in research and development, but also in marketing, management as well as the involvement of specialists in the different intellectual property rights around the globe [12].

Prof. Dr. Goto points out that the protection of intellectual property is an important issue to support the return on investment in research and development. Obtaining a patent should be seen as one of the many pieces of a holistic IP strategy. Patents are an important tool that allows intangible intellectual property developed by companies and universities to be protected and are the main form of protection of scientific work in general. For scientists and engineers it is therefore important to know what constitutes a patentable invention and to have a general idea about the procedures to apply for a patent.

5. The international context of patents

The internationalisation of IP rights started among others with the Paris convention (1883) followed by a number of treaties and agreements, which have been developed during a time span of more than 100 years. Although the international Patent Cooperation Treaty was established in 1970, it took another 24 years to arrive at the General Agreement on Tariffs and Trade (GATT) in 1994, which comprises the Agreement on Trade related Aspects of Intellectual Property Rights (TRIPS).

Independent from the TRIPS agreement and about two decades earlier, the first regional patent organization was founded based on the European Patent Convention (EPC), which was signed in 1973 and entered into force in 1977 [13]-[15]. Since this mile stone, an increasing regionalisation of intellectual property organizations has taken place throughout the world: the African Regional Intellectual Property Organization (ARIPO) in 1976, its French counterpart the 'Organisation Africaine de la Propriété Intellectuelle (OAPI) in 1977, the Eurasian Patent Convention (EAPC) in 1995 and the Andean Common Intellectual Property Regime (2000). The process of regionalisation appears to be far from finished. Indeed, it can be pointed out that in 1995 the countries of South-East Asia agreed on the ASEAN Patent Cooperation Treaty and more recently, in 2011, the South American IP Cooperation Project was initiated. Looking at this on-going process of regionalisation of IP it appears that these regional IP organisations are almost identical copies of the EPC system. It is not surprising that the EPC has become a model for regional IP organisations. Although each country in Europe has its own system for IP rights, for a uniform approach on patent rights, 38 European countries agreed to work together under the European Patent Convention (EPC) (Figure 3). Thanks to the powerful combined effort of these European countries the European Patent Office (EPO) has quickly become one of the leading patent offices in the world. Additionally a large amount of international applications filed under the regulation of the Patent Cooperation Treaty are in fact treated by the EPO as one of WIPO's International Search Authorities or International Preliminary Examination Authorities.

International cooperation is also increasing at the level of the classification of patent documents. The EPO recently announced on its website [16] that '*the Cooperative Patent Classification (CPC) is the most refined classification system for patent documents in the world (250 000 subdivisions) and was jointly implemented by the EPO and the United States Patent and Trademark Office on 1 January 2013. It is largely based on the European Classification (ECLA) system formerly used at the EPO. The CPC is already used by more than 45 patent offices worldwide as a means to perform efficient prior art searches during the patent granting process'*. The State Intellectual Property Office of the People's Republic of China (SIPO), the Federal Service for Intellectual Property of the Russian Federation (Rospatent) and the Eurasian Patent Office (EAPO) also decided to join this cooperative effort.

For Japanese researchers and engineers it is thus more and more important to become familiar with the international aspects of patent rights in view of this ongoing and increasing regional and international cooperation. More specifically, it is important to understand the particularities of the EPC and the role played by the EPC in this process of regionalisation and globalisation of IP rights.

6. Procedural aspects of filing a patent application

Each region or country has its own specific legal and procedural system. However, each patent system is characterized by general common principles concerning the procedure of applying for a patent. The two important phases in the procedure are, firstly, the investigation of the 'prior art', mostly scientific literature published before the date of filing the patent application, and, secondly, an examination of the patent application by a patent examiner in view of relevant prior art to establish conformity with the local patent law system. There are two major criteria, which determine what is a patentable invention: the invention must be novel and must involve an inventive step. These criteria are applied quite similarly throughout the global patent system. Finally most patent systems provide the possibility to appeal a refusal of a patent application or for third parties to oppose a granted patent.

The international Patent Cooperation Treaty procedure of filing a patent application at the World Intellectual Property Organization (WIPO) is usually followed by entering the regional (e.g. EP) and/or national (e.g. JP or US) patent system. In the lectures at the University of Tsukuba the procedure for filing an international and European patent application before the European Patent Office is used to illustrate the general principles of filing a patent application.

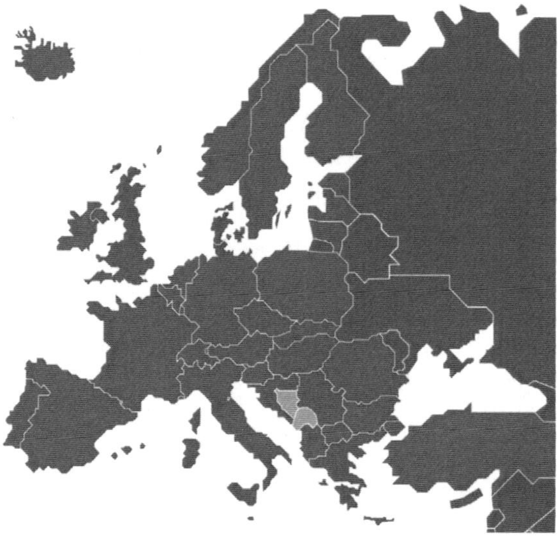

Figure 3. The 38 countries working together under the European Patent Convention.

7. The European Patent Office and the European patent grant procedure

Although each country in Europe has its own system for IP rights, for a uniform approach on patent rights, 38 European countries agreed to work together under the EPC (Figure 3). For the EPO to grant a European patent in respect of an invention, its description and claims have to meet certain criteria. These are laid down in the EPC [13] and supported by the Guidelines for Examination [14], the latter being developed on the basis of the case law of the boards of appeal, the EPO's second instance in-house judiciary. The most important criteria to be met are the requirements that the invention, as defined in the claims of the patent application, is novel and involves an inventive step (Articles 54 and 56 EPC). These criteria are evaluated against the 'prior art', which consists of all knowledge made available to the public before the filing date of the patent application, commonly in the form of a publication. The evaluation ('examination') of the technical criteria is carried out in writing by an 'examining division' consisting of three technically qualified examiners (Article 18 EPC). If the examining division intends to refuse a patent application, the procedure is concluded by a non-public hearing ('oral proceedings'), which involves a highly technical assessment of the invention in view of

IOP Conf. Series: Materials Science and Engineering **54** (2014) 012022 doi:10.1088/1757-899X/54/1/012022

scientific and technical literature to determine whether the patent application meets the EPC requirements. At the hearing the owner of the invention, also called 'applicant' or 'proprietor', is usually represented by a patent attorney.

The European patent grant procedure is an examination procedure beginning with a formalities examination and a mandatory documentary search. The first stage ends with the publication of the European patent application and the search report. At the applicant's request this is followed by the second stage, substantive examination. After the patent has been granted, there may be a further procedure in the form of opposition proceedings or, upon request of the patentee, limitation or revocation proceedings.

On receiving an application the EPO examines whether it can be accorded a date of filing. This is the case if the application documents contain: an indication that a European patent is sought; information identifying the applicant; a description or a reference to a previously filed application. If the application has been accorded a date of filing and is not deemed to be withdrawn, the Receiving Section checks for compliance with the provisions governing translations, the content of the Request for Grant, the presence of claims, the filing of the abstract, representation, formal requirements, any priority claimed, designation of the inventor and the filing of any drawings.

While the formalities examination is in progress, the European search is performed. The search report is drawn up on the basis of the claims, with due regard to the description and any drawings. It mentions the documents available to the EPO when it is drawn up which may be taken into consideration in assessing novelty and inventive step. The search report is accompanied by an opinion on whether the application and the invention to which it relates meet the requirements of the EPC. The non-binding opinion is not published together with the search report but is available to the public by way of file inspection after publication of the application.

Once the request for examination has been filed, the EPO examines, in the light of the search report and the applicant's response to it, whether the application and the invention to which it relates meet the requirements of the Convention, and in particular whether the invention is patentable. If the application and the invention to which it relates meet the requirements of the Convention, the examining division will decide to grant a European patent provided that the requisite fees have been paid in due time and a translation of the claims into the other two official languages of the EPO has been filed in due time. The grant does not take effect until the date on which it is mentioned in the European Patent Bulletin. At the same time as it publishes this mention, the EPO publishes a European patent specification containing the description, the claims and any drawings. The European Patent Bulletin is published electronically on the EPO 's publication server (www.epo.org).

Up to nine months after publication of the mention that a European patent has been granted, anyone may give the EPO notice of opposition to the patent. As a result of the examination of the opposition the opposition can be rejected, or the patent can be maintained in amended form or be revoked.

Appeals may be filed against decisions of the Receiving Section, the examining divisions, the opposition divisions and the Legal Division. An appeal has suspensive effect, which means that the contested decision is not yet final (no formal res judicata) and its effects are suspended. [15]

8. Conclusion and future prospect

In addition to the acquisition of scientific knowledge and research experience, it is also important that the education of researchers and engineers includes an important focus on developing knowledge about intellectual property rights. Because researchers and engineers play a key role as inventors, they need to become aware of their role in the patent application process. This knowledge should not be limited to the technical and legal aspects of patent law systems, but should also include an awareness of the international regionalisation processes which are increasingly influencing the acquisition of IP rights at the global level.

The lectures at the University of Tsukuba combine essential theoretical knowledge with practical group work. The experience from straightforward examples is at a later stage applied to the personal research topics of the students. As there are a variety of research groups at the Institute of Materials Science, the students are provided with the opportunity to discuss various research topics such as conducting polymers, solid state electronic devices and biological processes. During the presentations of their group work students experience and realize how the same principles of patentability are applied in various topics of research and technology.

The aim is to continue both educational activities for undergraduate, graduate and PhD students on the international aspects of intellectual property rights with a focus on patents, as well as educational research using synthesis of conjugated polymers. The didactical approach aims at merging student research with intellectual property education and can therefore be considered as unique in its synergistic approach to promoting simultaneously practical scientific research and knowledge on intellectual property. It is believed that the students having both a profound knowledge of intellectual property rights such as patents in an international context and a thorough experience with carrying out scientific research will be well equipped to take an active role in the community of science and technology after their graduation.

9. Acknowledgments

The views expressed here are exclusively those of the authors and do not reflect the position of the European Patent Office on the subject-matter of the article. We thank the European Patent Office, the Division of Materials Science (University of Tsukuba), and the Tsukuba Research Center for Interdisciplinary Materials Science (TIMS) for their support. The constructive feedback from Rainer Osterwalder, Jeremy Philpott, Irene Schellner and Marco Mergoni is highly appreciated.

References

[1] EPO-JPO Bilateral exchange project Nr. BEE02/2009
[2] http://www.cpos.ucsb.edu/, last modified 06-05-2010, accessed 30-09-2013
[3] Goto H; Akagi K and Shirakawa H 1997 *Synth Met* **84** 385
[4] Akagi K; Goto H; Okuda M and Shirakawa H 1998 *Mol Cryst Liq Crys.* **316** 201
[5] Chen W C and Jenekhe S A 1995 *Macromolecules* **28** 454
[6] Chen W C and Jenekhe S A 1995 *Macromolecules* **28** 465
[7] Kiebooms R and Wudl F 1999 *Synth Met* **101** 40-43
[8] Kiebooms R, Goto H and Akagi K 2001 *Macromolecules* **34 (23)** 7989–7998
[9] Innami Y, Kawashima H, Kiebooms R, Aizawa H, Matsuishi K, and Goto H 2012 *Materials* (Swiss) **5**, 317
[10] Innami Y, Kiebooms R H L, Koyano T, Ichinohe M, Ohkawa S, Kawabata K, Kawamatsu M, Matsuishi K, and Goto H 2011 *J Mater Sci* **46** 6556
[11] Schmeisser W, Mohnkopf H, Hartmann M, and Metze G 2009 *Innovation performance accounting: Financing Decisions and Risk Assessment of Innovation Processes* (Berlin: Springer Verlag)
[12] Palfry J 2012 *Intellectual Property Strategy* (Cambridge, Massachusetts: The MIT Press)
[13] *The European Patent Convention* 2010 14th edition, European Patent Office, http://www.epo.org/law-practice/legal-texts/epc.html accessed 26-09-2013
[14] *Guidelines for examination in the European Patent Office* 2013 European Patent Office, http://www.epo.org/law-practice/legal-texts/guidelines.html accessed 26-09-2013
[15] *How to get a European patent - Guide for applicants* 2010 13th edition, European Patent Office, http://www.epo.org/applying/european/Guide-for-applicants.html accessed 26-09-2013
[16] *Co-operation with Russia and Eurasian region intensifies: Language barrier in patent information removed.* 2013 European Patent Office http://www.epo.org/news-issues/news/2013/20130926.html, accessed 02-10-2013, last updated 26-09-2013

IOP Conf. Series: Materials Science and Engineering **54** (2014) 012023 doi:10.1088/1757-899X/54/1/012023

Solubility control of regioregular 3-substituted polythiophenes bearing 2-phenylnaphthalene side chain by copolymerisation

Mari Watanabe[1,2] and Masashi Kijima[2,3]

[1] Graduate School of Pure and Applied Sciences, University of Tsukuba, Tennodai 1-1-1, Tsukuba, Ibaraki 305-8573, Japan
[2] Tsukuba Research Center for Interdisciplinary Materials Science
[3] Faculty of Pure and Applied Sciences, University of Tsukuba

E-mail: kijima@ims.tsukuba.ac.jp

Abstract. Two types of 2,5-dibromothiophene monomers having a dodecyl chain and a 2-phenylnaphthalene one were randomly copolymerised in different feed molar ratio by Ni-catalysed chain-growth polymerisation method. Regioregularity of all polymers were sufficiently high (88-97%). Absorption λ_{max} due to $\pi-\pi^*$ transition of polythiophene backbones in solutions were observed at 444-451 nm for all polymers, which were almost similar to each other and typical of the optical characteristics of regioregular poly(3-alkylthiophene)s. The homopolymer of the latter monomer that had the phenylnaphthalene side group showed poor solubility to common organic solvents, i.e., it was insoluble in chloroform at room temperature and could dissolve only in hot solvents above 100 °C. The copolymers, which had higher number average molecular weights (M_n), had better solubility than that of the homopolymer.

1. Introduction

Polythiophene and related derivatives having alkyl side chains have been studied worldwide energetically, while examples of studies about polythiophene derivatives having aromatic side groups with special functionalities such as liquid crystalline properties are limited so far. Recently, our group realised synthesising 3-substituted polythiophenes bearing a liquid crystalline and semiconducting 2-phenylnaphthalene moiety [1]. Polymers provided by the same 2,5-dibromo-3-substituted thiophene monomer but different polymerisation methods exhibited quite different properties. The polymers prepared by Yamamoto dehalogenative polycondensation method showed modest head-to-tail regioregularity, good solubility in organic solvents, large liquid crystalline domains, and wide energy gap (E_g) between HOMO and LUMO levels. In contrast, the polymers prepared by Ni-catalysed chain-growth polymerisation method showed better regioregularity, poor solubility, smaller liquid crystalline domains, and narrower energy gap. The latter polymers must have unique semiconducting properties of the regioregular poly(3-substituted thiophene) backbone and aggregated phenylnaphthalene side chain in an oriented state, however, the unexpected low solubility due to large intermolecular interaction holds up further investigations. In this study, 3-substituted thiophene bearing 2-phenylnaphthalene moiety was copolymerised with 3-alkylthiophene by the Ni-catalysed chain-growth polymerisation method in order to control solubility and regioregularity of poly(3-substituted thiophene)s bearing 2-phenylnaphthalene side chains, and basic properties of these polymers were investigated.

2. Experimental

2.1. Materials

One of the monomers, 2,5-dibromo-3-dodecylthiophene (**M1**), was synthesised along with well-established procedure from 3-bromothiophene [2,3], and the other monomer, 2,5-dibromo-3-{10-[6-(4-octylphenyl)naphthalene-2-yloxy]decyl}thiophene (**M2**), was prepared by the procedures described in our previous report [1]. Other chemicals used in this study were purchased from Kanto Chemical, Tokyo Chemical Industry, Nacalai Tesque, Sigma-Aldrich, and Acros Organics.

2.2. General procedure for the copolymerisation

All polymers were synthesised by Ni-catalysed chain-growth polymerisation method [4, 5]. The feed molar ratios of **M1**:**M2** were settled as 1:0, 1:1, 1:3, 1:5, and 0:1, and resultant polymers were abbreviated $\mathbf{P}_{1:0}$, $\mathbf{P}_{1:1}$, $\mathbf{P}_{1:3}$, $\mathbf{P}_{1:5}$, and $\mathbf{P}_{0:1}$, respectively (Figure 1). Polymerisation was carried out under the conditions in the presence of the monomers (0.1 M (mol dm^{-3})), 2.0 M isopropylmagnesium chloride (*i*PrMgCl) in THF (1.05 eq), a Ni catalyst (0.02 eq), and LiCl (1-2 eq) in dry THF and under a N_2 atmosphere.

Typical procedure was; the monomers (**M1, M2)** and LiCl were dissolved in dry THF. To the mixture, *i*PrMgCl (2.0 M) in THF was added at 0 °C, which was stirred at 0 °C for 1 h. After addition of [1,3-bis(diphenylphosphino)propane]dichloronickel(II) (Ni(dppp)Cl_2), an ice bath was removed and the reaction mixture was warmed up to room temperature slowly. After reacting 1-12h, the reaction was quenched by adding an aqueous HCl (5 M). The polymers were purified by precipitation from MeOH and acetone. For $\mathbf{P}_{1:1}$, $\mathbf{P}_{1:3}$, and $\mathbf{P}_{1:5}$, additional reprecipitation from hexane was performed. Similarly, $\mathbf{P}_{0:1}$ was further purified by reprecipitation from acetone/chloroform. Purple solids (40-62% yield) were obtained in all cases.

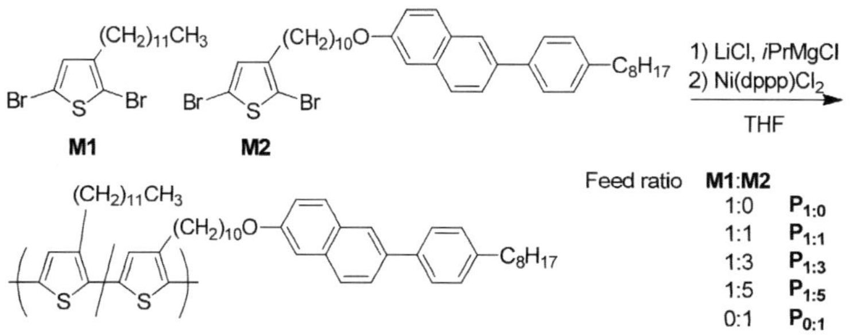

Figure 1. Synthetic pathways and structures of the polymers.

3. Results and discussion

3.1. Synthesis and characterisations of polymers

M1 and **M2** in the feed molar ratios of 1:0, 1:1, 1:3, 1:5, and 0:1 were randomly polymerised by the Ni-catalysed chain-growth method [4,5] to obtain regioregular poly(3-alkylthiophene)s ($\mathbf{P}_{1:0}$, $\mathbf{P}_{1:1}$, $\mathbf{P}_{1:3}$, $\mathbf{P}_{1:5}$, and $\mathbf{P}_{0:1}$) according a pathway shown in Figure 1. The number average molecular weights (M_n), weight average molecular weights (M_w), polydispersity indexes (M_w/M_n), number average degrees of polymerisation (dp), regioregularities, and actual compositon ratios of polymers were listed in Table 1. To determine M_n and dp of $\mathbf{P}_{0:1}$, ^{1}H NMR spectrum in hot chlorobenzene-d5 was used because of its poor solubility to the gel permeation chromatography (GPC) eluent (THF) at ambient temperature [1]. Figure 2 shows ^{1}H NMR spectra of the polymers. Average molecular weights of the polymers except $\mathbf{P}_{0:1}$, which were determined by GPC, were enough high and similar to those of poly(3-alkylthiophene)s provided by a similar method [6]. Average formula masses considered actual composition ratios of two units were used to calculate dp. The actual composition ratios of the copolymers were calculated from integration ratio between a well-separated ^{1}H NMR peak at 4.0 ppm due to naphthoxy-linked methylene protons of the **M2** unit (D) and overlapped peaks at 0.9 ppm due to terminal methyl groups of both monomer units (B and F). It is verified that reactivities of two monomers during polymerisation were almost same, because composition molar ratios of the polymers were quite similar with their initial feed molar ratios of the monomers. Regioregularities of the polymers were estimated similarly by the method in our previous report [1] considering integration values at methylenes A, C, and E in the ^{1}H NMR spectra (Figure 2), and revealed to be high and similar each other except for $\mathbf{P}_{0:1}$. The relatively low regioregularity observed for $\mathbf{P}_{0:1}$ was attributed to its low molecular weight.

Table 1. Polymerisation results.

Polymer	M_n / kg mol^{-1}	M_w[a] / kg mol^{-1}	M_w/M_n[a]	dp[c]	Regioregularity[d] / %	Composition ratio[d] **M1**:**M2**
$\mathbf{P}_{1:0}$	11.0[a]	14.8	1.34	43.9[a]	97	1:0
$\mathbf{P}_{1:1}$	17.9[a]	21.8	1.22	42.1[a]	97	1:1.1
$\mathbf{P}_{1:3}$	15.9[a]	21.0	1.32	33.0[a]	96	1:3.4
$\mathbf{P}_{1:5}$	18.5[a]	23.4	1.26	37.0[a]	92	1:4.9
$\mathbf{P}_{0:1}$	(4.09)[b]			(7.4)[b]	88	0:1

[a] Determined by GPC calibrated with polystyrene standards using THF as an eluent.
[b] Estimated by ^{1}H NMR spectrum [1].
[c] Number-average degree of polymerisation.
[d] Determined by ^{1}H NMR spectra.

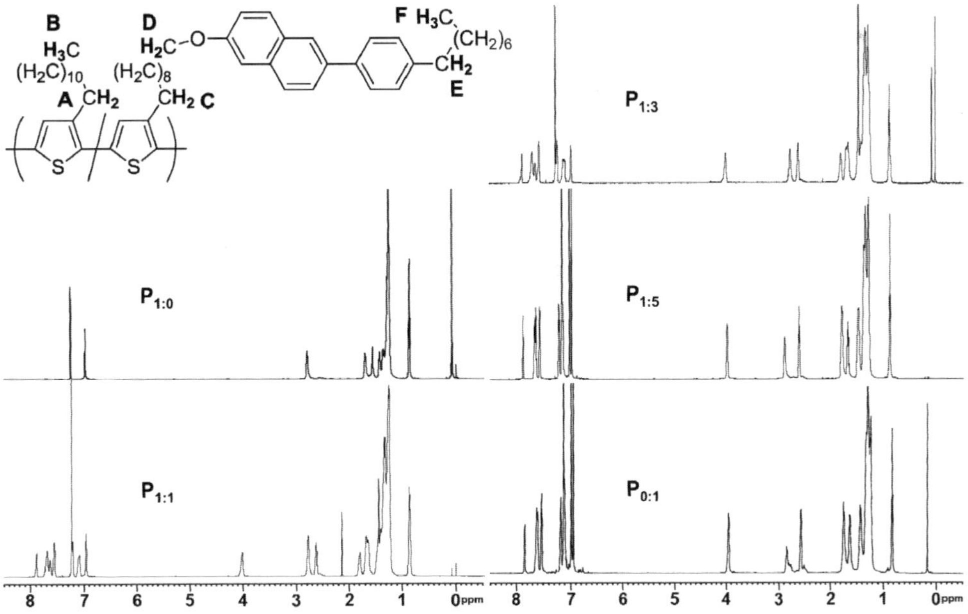

Figure 2. ^{1}H NMR spectra of $\mathbf{P}_{1:0}$ in $CDCl_3$ at rt, $\mathbf{P}_{1:1}$ in $CDCl_3$ at 323K, $\mathbf{P}_{1:3}$ in $CDCl_3$ at 323K, $\mathbf{P}_{1:5}$ in chlorobenzene-d_5 at 373K, and $\mathbf{P}_{0:1}$ in chlorobenzene-d_5 at 393K at a resonance frequency of 600 MHz.

3.2. Solubilities of polymers

Solubilities of the polymers were evaluated using chloroform which is popular and most often-used good solvent for conjugated polymers. Figure 3 is pictures of the polymer solutions in chloroform. Weight of polymer and volume of chloroform were determined 1.0 mg and 1.0 mL. All the polymers were purple in solid state but their colour changed to light orange when they were fully dissolved. After well mixing them at respective temperatures of ambient, 40 °C, 50 °C, and 60 °C, the mixing samples were left to stand at least 5 min at each temperature to check solubility. The homopolymer of **M1**, $\mathbf{P_{1:0}}$, was soluble in chloroform at room temperature as is well known. On the other hand, the homopolymer of **M2**, $\mathbf{P_{0:1}}$, did not dissolve even if it was heated at 60 °C in spite of its low molecular weight. Solubility of the copolymers was increased as increasing the content ratio of **M1** unit, that is, $\mathbf{P_{1:5}}$, $\mathbf{P_{1:3}}$, and $\mathbf{P_{1:1}}$, dissolved completely at above 60 °C, 40 °C, and room temperature, respectively. From these results, it was found that the copolymerisation with 3-dodecylthiophene was effective for improving solubility of regioregular poly(3-substituted thiophene) bearing 2-phenylnaphthalene side group. It is suggested that the 3-dodecylthiophene (**M1**) unit encouraged dissolving and disturbing aggregation of the phenylnaphthalene side group. Differential scanning calorimetry (DSC) measurement at a heating/cooling rate of 10 °C/min was used to investigate phase transition behaviours of the polymers. Because of their high intermolecular forces in condensed states, the transition temperatures from isotropic liquid to the first less mobile phase of $\mathbf{P_{1:0}}$, $\mathbf{P_{1:1}}$, $\mathbf{P_{1:3}}$, $\mathbf{P_{1:5}}$, and $\mathbf{P_{0:1}}$ were clearly observed at 115 °C, 107 °C, 114 °C, 164 °C, and 171 °C, in the second cooling process of DSC thermograms. Although these phase trtansition temperatures are affected by molecular weights of polymers, it can be said that introduction of 3-dodecylthiophene (**M1**) unit was effective to improve fusibility of poly(3-substituted thiophene) bearing 2-phenylnaphthalen side group, $\mathbf{P_{0:1}}$.

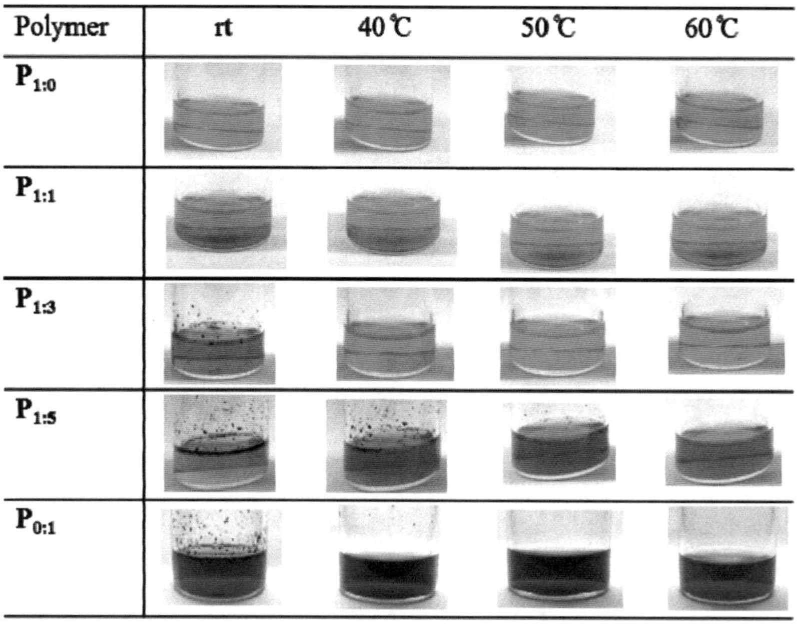

Figure 3. Pictures of polymer (1.0 mg) solutions in chloroform (1.0 mL).

3.3. Optical properties

UV-vis absorption spectra of the polymers in solution states and drop-cast films were shown in Figure 4 and Figure 5, and data of the UV-vis absorption λ_{max} and photoluminescence (PL) λ_{em} of the polymers in solution and film states due to π-π* transition were listed in Table 2. The absorption λ_{max} and PL λ_{max} of the polymers in solutions were observed at 444-451 nm and 579-588 nm, and they were similar each other and almost identical to the data of regioregular poly(3-dodecylthiophene) in a previous report [7]. On the other hand, the absorption λ_{max} of $\mathbf{P_{1:1}}$, $\mathbf{P_{1:3}}$, and $\mathbf{P_{1:5}}$, in films were hypsochromic shifted than that of $\mathbf{P_{0:1}}$, which might be due to effect of steric hindrance of the bulkier phenylnaphthalene side group than the dodecyl group (Figure 5). However, absorption edges of $\mathbf{P_{1:1}}$, $\mathbf{P_{1:3}}$, and $\mathbf{P_{1:5}}$, in films were little shifted from that of $\mathbf{P_{0:1}}$, which means that these copolymers have almost same bandgaps (E_g) as $\mathbf{P_{0:1}}$ has.

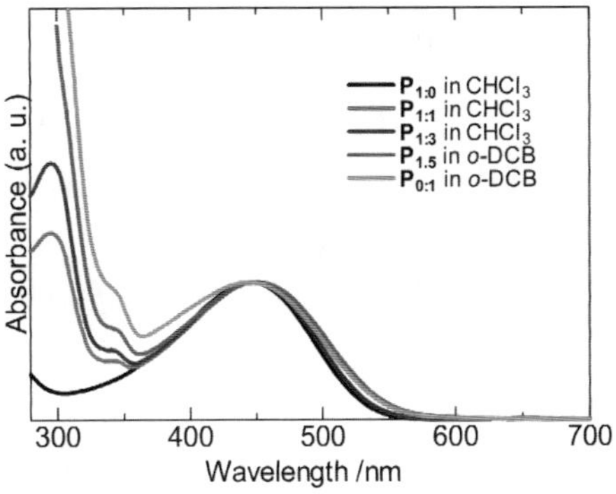

Figure 4. UV-vis absorption spectra of $\mathbf{P_{1:0}}$ (black), $\mathbf{P_{1:1}}$ (red), and $\mathbf{P_{1:3}}$ (blue) in chloroform, and $\mathbf{P_{1:5}}$ (green), and $\mathbf{P_{0:1}}$ (orange) in *o*-dichlorobenzene.

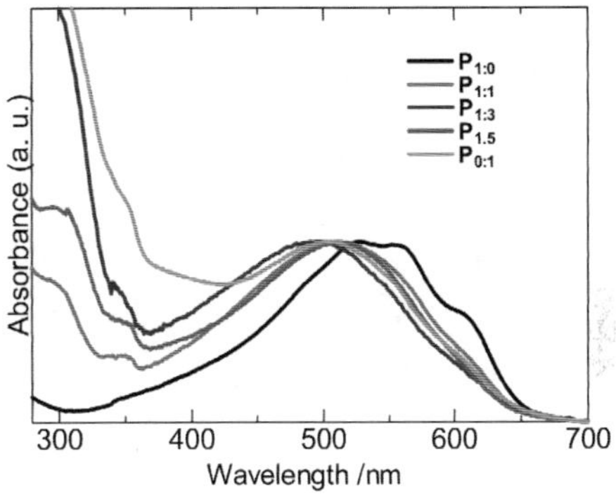

Figure 5. UV-vis absorption spectra of drop-cast films of $\mathbf{P_{1:0}}$ (black), $\mathbf{P_{1:1}}$ (red), $\mathbf{P_{1:3}}$ (blue), $\mathbf{P_{1:5}}$ (green), and $\mathbf{P_{0:1}}$ (orange).

Table 2. Optical properties of the polymers.

Polymer	UV-vis λ_{max}/nm[a]		E_g/eV	PL λ_{em}/nm[a]	
	Solution	Film		Solution	Film
$\mathbf{P_{1:0}}$	445[b]	527,553	1.94	579[b]	727
$\mathbf{P_{1:1}}$	451[b]	512	1.95	580[b]	692
$\mathbf{P_{1:3}}$	447[b]	494	1.95	580[b]	628
$\mathbf{P_{1:5}}$	451[c]	511	1.96	588[c]	714
$\mathbf{P_{0:1}}$	444[c]	509	1.95	585[c]	720

[a] UV-vis λ_{max} and PL λ_{max} due to π−π* transition of polythiophene backbone.
[b] In chloroform.
[c] In *o*-dichlorobenzene.

3.4. Energy levels

Energy levels of highest occupied molecular orbital (E_{HOMO}) of the polymers were estimated from the onset oxidation potentials of the polymers measured by cyclic voltammetry (CV), and Figure 6 shows the oxidation waves of the polymers. E_{HOMO} values of $\mathbf{P_{1:0}}$, $\mathbf{P_{1:1}}$, $\mathbf{P_{1:3}}$, $\mathbf{P_{1:5}}$, and $\mathbf{P_{0:1}}$ estimated from each onset potential were −4.91, −5.18, −5.24, −5.25, and −5.40 eV, although these polymers had almost same E_g. The E_{HOMO} of $\mathbf{P_{0:1}}$ was deeper than that of $\mathbf{P_{1:0}}$ and the E_{HOMO} values of the copolymers were located between the homopolymers. This phenomenon implies that steric hindrance and/or some specific electronic interaction exist between the polythiophene backbone and the 2-phenylnaphthalene side group, which stabilises polythiophene backbone against oxidisation.

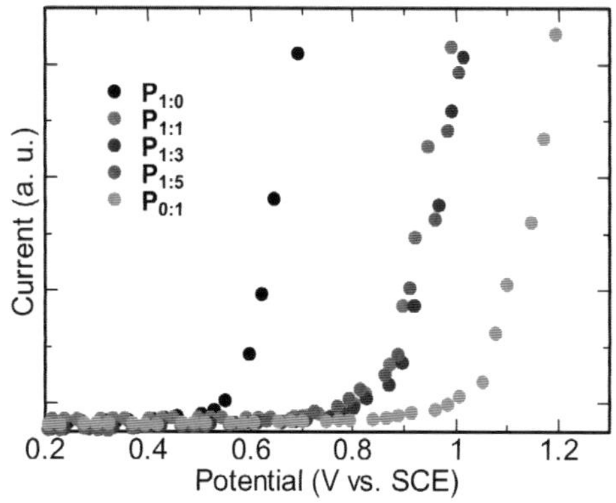

Figure 6. Difference of onset potentials in oxidation waves of the polymers observed by CV in 0.1 M Et_4NBF_4/acetonitrile at a scan rate of 50 mV/s.

4. Conclusions

A thiophene monomer, 2,5-dibromo-3-{10-[6-(4-octylphenyl)naphthalene-2-yloxy]decyl}thiophene (**M2**), was randomly copolymerised with 2,5-dibromo-3-dodecylthiophene (**M1**) by the Ni-catalysed chain growth method in order to improve solubility of regioregular poly(3-substitutedthiophene) with 2-phenylnaphthalene side group. The copolymers, especially $\mathbf{P_{1:1}}$, realised much better solubility in organic solvents than $\mathbf{P_{0:1}}$ with maintaining high regioregularity. It is proved that insertion of the 3-dodecylthiophene moiety into a homopolymer structure of **M2** can disturb aggregation between the 2-phenylnapthalene moieties and/or polythiophene backbone, effectively. In contrast, it is assumed that presence of the 2-phenylnaphthalene moiety in the regioregular 3-substituted polythiophenes weakens π-stacking interaction between polythiophene backbones in consequence of steric effect of the bulky side chain in some degree, which lowered their E_{HOMO} levels in spite of bearing similar regioregularity and E_g.

Acknowledgements

The authors thank Prof. H. Goto, University of Tsukuba, for the mesurement of GPC, and Chemical Analysis Division, Research Facility Center for Science and Technology, University of Tsukuba, for facilities of the NMR, elemental analysis, DSC, UV–vis, and PL measurements.

References

[1] Watanabe M, Tsuchiya K, Shinnai T, and Kijima M 2012 *Macromolecules* **45** 1825
[2] Pham C V, Mark Jr. H B, and Zimmer H 1986 *Synth. Commun.* **16** 689
[3] Lanni E L, and McNeil A J 2009 *J. Am. Chem. Soc.* **131** 16573
[4] Iovu M C, Sheina E E, Gil R R, and McCullough R D 2005 *Macromolecules* **38** 8649
[5] Miyakoshi R, Yokoyama A, and Yokozawa T 2005 *J. Am. Chem. Soc.* **127** 17542
[6] Ge J, He M, Qiu F, and Yang Y 2010 *Macromolecules* **43** 6422
[7] Chen T A, Wu X, and Rieke R D 1995 *J. Am. Chem. Soc.* **117** 233

IOP Conf. Series: Materials Science and Engineering **54** (2014) 012024 doi:10.1088/1757-899X/54/1/012024

Novel function of polyaniline for biological environments: Cultivation of paramecium in the presence of polyaniline

Hiromasa Goto

Division of Materials Science, Faculty of Pure and Applied Sciences, University of Tsukuba, Tsukuba, Ibaraki 305-8573, Japan

E-mail: gotoh@ims.tsukuba.ac.jp

Abstract. The reduced form of polyaniline (PANI, emeraldine base) functions as water purification to extend the lives of paramecia. The emeraldine base can absorb discharged waste from the planktons such as nitrogenous compounds and salts. This is a new function of π-conjugated polymers for micro-organisms.

1. Introduction

Leuwenhoek (the Netherlands) firstly observed paramecium under optical microscopy [1]. Movement of the paramecium in water is by its cilia. In the closed environment, such as water in a small beaker, paramecium cannot live for long. Increase of population of paramecia in a beaker is observable at first, however, at some point the population suddenly decreases because of discharged wastes, such as nitrogenous compounds and salts. The water waste from paramecia decimates the paramecium population itself in such the restricted circumstance. Planktons have no ornithine cycle, so they cannot convert ammonia to nontoxic urea. The discharged ammonia pollutes their living environment. Fortunately, in the natural realm, the circulation system detoxifies unwanted materials and provides good conditions for living organisms.

Polyaniline (PANI) is one of the most promising conducting polymers because it possesses relatively high electrical conductivity, corrosion resistance [2], and antistatic functions [3]. Recently, radioactive-atom adsorption function of PANI for air cleaning system has been proposed [4]. The emeraldine base (EB) of the PANI can absorb electron acceptors (acids) by the electron donor-acceptor (dopant) interaction. This research explores a new function of PANI for collection of pollutants.

In this report, cultivation of paramecia in the presence of PANI emeraldine base (PANI-EB, half-doped form) is examined.

2.Paramecium

Three kinds of paramecia, G3 paramecium caudatum, $Sol_5 5$/white paramecium bursaria, and $Sol_5 5$/green paramecium bursaria were employed in this study. A small amount of chlorella is placed in the water for food for the paramecia.

2. Synthesis

2.1. Polyaniline

A solution of aniline (1 g) and sulfuric acid (1 g) in water (200 mL) was stirred for 1 h. The solution was cooled by ice. Then, ammonium peroxodisulfate (1.5 g) was slowly added to the solution. After 24 h, the solution was filtered. The crude PANI cake on the funnel was collected, and washed with a large volume of water (1000 mL). The solution was filtered again. The solid was collected and washed with a large volume of methanol (200 mL). After a further filtration, the solid was washed with a large volume of ammonia/water. The colour of PANI changed from dark green to dark blue upon the treatment with the ammonia solution (Scheme 1). The product was collected after filtration, and dried in vacuum to yield PANI-EB. PANI before treatment with ammonia was also obtained as PANI emeraldine salt (PANI-ES, doped form).

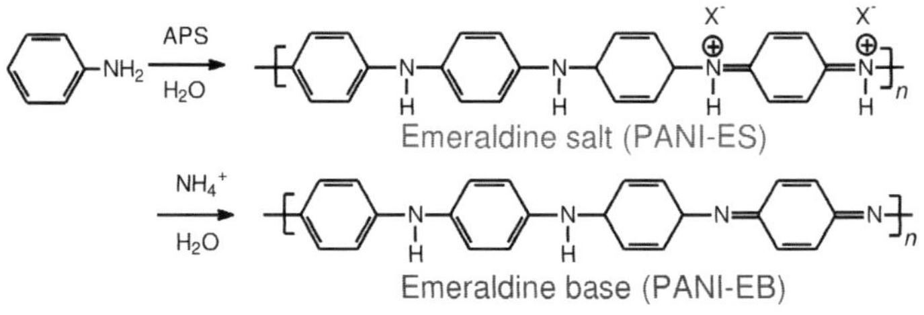

Scheme 1. Synthesis of polyaniline emeraldine salt (PANI-ES) and emeraldine base (PANI-EB). APS = ammonium persulfate.

2.2. Poly(pyridine-vinylene-azobenzen), P(Py-Vy-Az)

The copolymerisation was carried out by the Migita-Kosugi-Stille coupling polycondensation followed by Mizoroki-Heck reaction with the one-pot synthesis method. Synthetic route for poly(pyridine-vinylene-azobenzen) (P(Py-Vy-Az)) is shown in Scheme 2. *p*, *p*'-Dibromoazobenzene ($AzBr_2$) was prepared from *p*-bromonitrobenzene by using MnO_2, according to the literatures [5,6]. A solution of 2,6-dibromopyridine (0.3 g, 1.28 mmmol), [tris(dibenzylideneacetone)dipalladium(0)] ($Pd(dba)_3$, 22 mg, 0.24 mmol), tris(*o*-furyl)phosphine ($P(fulyl)_3$, 24 mg, 0.24 mmol), tributyl(vinyl)tin (0.8 g, 2.5 mmol) in tetrahydrofuran (THF, 1 mL) was refluxed for 24 h at 60 °C. Another solution of *p*, *p*'-dibromoazobenzene (0.43g, 1.28 mmol), palladium diacetate (II) [$Pd(OAC)_2$] (56 mg, 0.25 mmol), tris(*o*-tolyl)phosphine ($P(o\text{-tol})_3$, 0.4 g, 1.3 mmol) in *N*,*N*-dimethylformamide (DMF, 1 mL) was added and stirred for 3 days at 60 °C. Then, the reaction mixture was poured in a large volume of methanol (ca. 200 mL), and HCl was added to the solution to remove the catalysts. After filtration, the product was washed with methanol containing ammonia/H_2O solution. The polymer was collected by filtration, and dried in vacuum to yield 0.52 g of dark red powder.

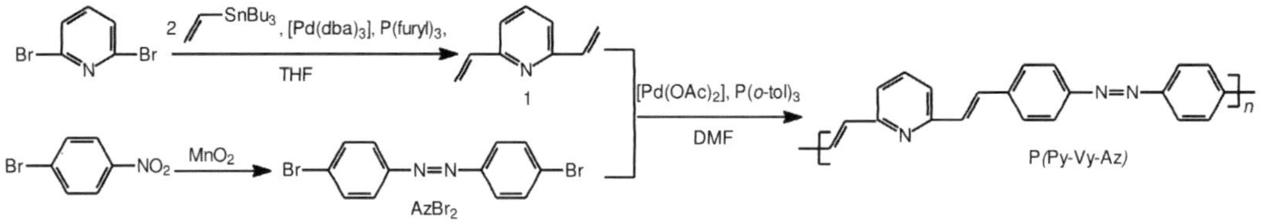

Scheme 2. Synthetic route for poly(pyridine-vinylene-azobenzene), P(Py-Vy-Az). $Pd(dba)_3$ = tris(dibenzylideneacetone)dipalladium(0), $P(fulyl)_3$ = tris(*o*-furyl)phosphine, THF = tetrahydrofuran, $Pd(OAC)_2$ = palladium diacetate (II), $P(o\text{-tol})_3$ = tris(*o*-tolyl)phosphine, Bu_3Sn = tetrabutyl tin, DMF = *N*,*N*-dimethylformamide.

3. Cultivation of paramecia

Paramecia in water were added to the water containing the PANI-EB or P(Py-Vy-Az). The PANI-EB and P(Py-Vy-Az) were insoluble in the water, so the fine particles of the polymers were dispersed.

The paramecia moved in the water in the presence of the polymers (Figure 1), although the paramecia could not live in the water containing the PANI-ES (full-doped form) because of strong acidity (pH < 4.3).

4. Observations

Oxidation-reduction (redox) potentials of water containing paramecia were measured as references. The oxidation-reduction potential (ORP) values were to be 281–348 mV (Table 1).

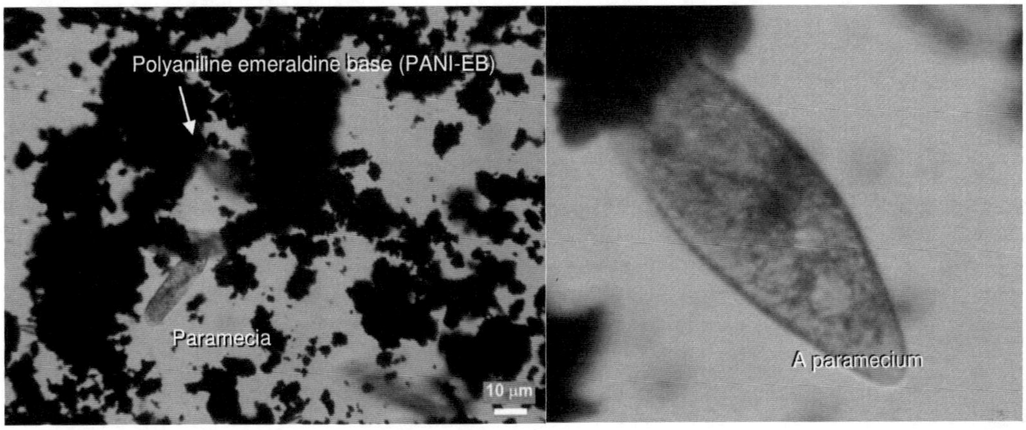

Figure 1. Paramecia with polyaniline emeraldine base (PANI-EB) in the water (left). A magnification image (right).

Table 1. ORP, pH measurement results.

Paramecium	Paramecium in water		PANI/paramecium in water	
	ORP[3] (mV)	pH	pH (initial)	pH (final)
G3 P. caudatum[1]	312	7.6	7.6	7.6
$Sol_5$5/white P. bursaria[1,2]	281	7.5	7.6	8.2
$Sol_5$5/green P. bursaria[1]	348	7.7	7.1	7.9

[1]P = paramecium.
[2]$Sol_5$5/white paramecium bursaria lived in the water containing P(Py-Vy-Az) for 10 days.
[3]ORP = oxidation-reduction potential.

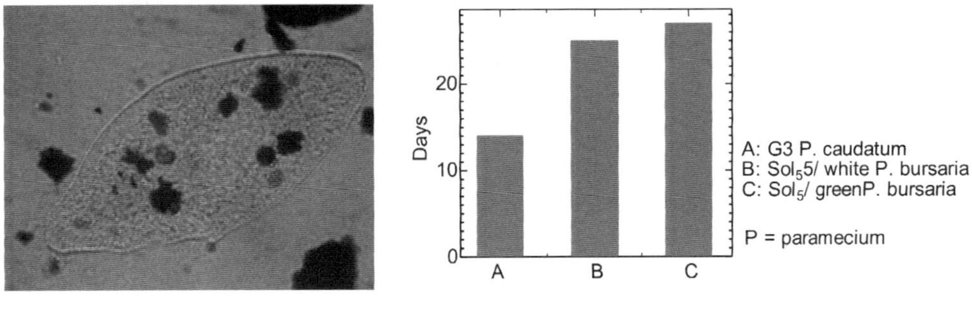

Figure 2. Differential interference optical microscopy image of a paramecium and PANI-EB in the water (left). Survival period of the paramecia in the presence of PANI-EB (right).

A differential interference optical microscopy image confirmed a figure of the paramecium (Figure 2, left). The paramecia in the flask could live for a long period in the water containing the PANI-EB (Figure 2, (right)). The PANI-EB is in a half-doped state. Therefore, the discharged wastes from the paramecia were neutralized by the PANI-ES. However, a paramecia in the water containing

P(Py-Vy-Az) lived for only 10 days. P(Py-Vy-Az) has no acid because it is the undoped form. The solution containing P(Py-Vy-Az) was not neutralized and the water was polluted by the material from the paramecia.

Discharged wastes from paramecia in the water were absorbed by the PANI-EB. The water purification function of the PANI-EB provides a good environment for long paramecium life. The pH value of the solutions are increased except for G3 P. caudatum, as shown in Table 1, indicating base substances such as ammonia from paramecia could be neutralized by the PANI-EB. Therefore, the PANI-EB provides a good environment for the paramecia.

5. Conclusions

At present, components of discharged wastes cannot be clearly identified because living organisms discharge many kinds of compounds. However, the longevity of the paramecium in the presence of the PANI-EB implies a water cleaning function of the PANI. Mechanical adsorption of discharged wastes by micro-pores of the PANI surface is also considerable. This is the first example of the application of polyanilines to the biological environment, and thus the possibility of π-conjugated polymer utility was further expanded.

Acknowledgment

Paramecia were generously provided by Prof. Mihoko Takahasi (Institute of Biological Sciences, U. Tsukuba). The author thanks Prof. Nobuhiko Nomura (Graduate School of Life and Environmental Sciences, University of Tsukuba) for valuable discussion and collaboration on polymer science and biotechnology.

References

[1] Palm L C 1982 *Antoni van Leeuwenhoek, 1632–1723: Studies on the Life and Work of the Delft Scientist Commemorating the 350th Anniversary of His Birthday* (Amsterdam: Rodopi BV)
[2] Chen X J, Luo C, Zhang Z H and Zhao M 2012 *Anti-Corrosion Methods and Materials*, **59** 291
[3] Skotheim T J, Elsenbaumer R L and Reynolds J R 1998 *Handbook of Conducting Polymers, 2nd Ed.*, 1097, (New York: Marcel Dekker)
[4] Goto H 2011 *Polymers* **3** 875
[5] Wheeler O H, Gonzalez D 1964 *Tetrahedron* **20** 189
[6] Sandler S R, Karo W 1971 *Organic Functional Group Preparations, Vol II* (New York and London: Academic Press) p321

Energy harvesting of dye-sensitized solar cells assisted with Ti-mesh and phosphor materials

Weizhen He, Atabaev Sh. Timur, Hyung-Kook Kim and Yoon-Hwae Hwang*

Department of Nanomaterials Engineering & BK 21 Nano Fusion Technology Division, Pusan National University, Miryang 627-706, South Korea

E-mail: yhwang@pusan.ac.kr

Abstract. We used the Ti-meshes for both the photoanode and counter electrode of the dye-sensitized solar cells (DSSCs) to improve the flexibility and conductivity of electrodes. This mesh type electrode showed good transparency and high bendability when subjected to an external force. The overall efficiency of the best cells was approximately 5.3% under standard air mass 1.5 global (AM 1.5 G) solar condition. We also modified the TiO_2 nanoparticle based dye-sensitized solar cells (DSSCs) by depositing a layer of long-persistent phosphor $SrAl_2O_4:Eu^{2+}$, Dy^{3+} on top of the TiO_2 nanoparticle layer to prepare working electrodes of DSSCs. $SrAl_2O_4:Eu^{2+}$, Dy^{3+} red-shifted short UV wavelengths into the main absorption range of the dye commonly used in DSSCs. The $SrAl_2O_4:Eu^{2+}$, Dy^{3+} layer also acted as a light-scattering layer to reduce the loss of visible light. An overall 13% improvement in the conversion efficiency of modified DSSCs was achieved due to the presence of the phosphor layer.

1. Introduction

Dye-sensitized solar cells (DSSCs) have been considered to be one of the most promising photovoltaic conversion technologies [1]. DSSCs are made from low-cost materials and they can be significantly less expensive than conventional solid-state solar cells. These cells can also be designed as flexible sheets and be mechanically robust [2]. Recently, plenty of innovations of the DSSC structure were reported [3-6]. Although a conversion efficiency of those innovative DSSC was less than the best thin-film cells, its price/performance ratio (kWh/(m^2·annum·dollar)) should be high enough to allow them to compete with traditional energy sources based electrical generation. DSSCs are still in their development stages, however, a higher efficiency gains are possible and more widespread studies have been carried out [2-13].

Considerable efforts have been devoted to the development of more efficient photoanode structures. In particular, 1-D TiO_2 nanotube (TNT) structures are attracting more interest from the solar cell community because of their superior charge transport speed, great diffusion length and small charge recombination rate [14,15]. The development of anodization method [16] makes it quite facile for the fabrication of uniform and highly ordered TNTs. On the other hand, this method requires the use of Ti foil which is not transparent and DSSCs based on anodic TNTs on Ti foil should be illuminated from the back side, leading to a decrease in light harvesting and hence low conversion efficiency. Also, the use of TCO glass as the counter electrode has restricted the efficiency to a low level ($\eta \sim 1.47\%$).

For DSSCs, efficiency relies on the light harvesting to a great extent, and the dye plays a key role in converting light to electrical power. However, the most commonly used Ru(II)

dyes in current DSSCs have an optical band gap of about 1.8 eV and only absorb portions of the UV and visible spectrum[17,18]. Thus, in order to achieve higher light harvesting, new-type dyes have been under development. A possible way to avoid the loss of photon energy evolved is the utilization of a luminescence layer composed of up-conversion (UC) or down-conversion (DC) phosphor materials[19-21]. DC phosphors have the ability to transform higher energy photons into visible photons, whereas UC phosphors work in the opposite way.

In this study, we present a DSSC with high flexibility utilizing Ti-mesh as both electrodes. An anodization method at a relatively low temperature was carried out to fabricate TNTs on a Ti-mesh. The TNTs on the mesh substrate were oriented in all directions ranging from parallel to perpendicular to the incoming photons. Furthermore, the voids in a mesh substrate allow light to enter the cell from the photoanode side leading to possible front illumination. We also describe DSSCs assisted by long-persistent $SrAl_2O_4:Eu^{2+}$, Dy^{3+} phosphor particles. Those have a wide UV-light absorption band and emit an output in the blue-green spectral regions, which matches the peak absorption region of the most commonly used dye N719. In addition, the micro-sized particles contribute to visible light scattering path length and hence, benefit light collection. Furthermore, the open-circuit voltage and the photocurrent of modified DSSCs were improved.

2. Experiments

For Ti-mesh based DSSC study, the meshes made from commercially pure titanium (99.5% purity) were purchased from Alfa Aesar. In this study, a 50-mesh wire cloth woven from 0.102 mm diameter wire with an open area of 64% was used. For an anodization, the mesh was fixed by contacting with a Cu plate as the anode and a graphite plate was used as the counter electrode. Anodization was performed at a constant potential of 60 V at 5 °C, and the growth time was varied from 4 h to 8 h. For phosphor particle based DSSC study, commercially-available P25 TiO_2 powder (Aldrich, 99.5% purity) was used and the mean particle size was approximately 21 nm. The $SrAl_2O_4:Eu^{2+}$, Dy^{3+} phosphor powder (99% purity, mean particle size: 5~10μm), purchased from Dalian Zhongbang Chemical Industry Co., Ltd., was mixed with 2 wt% ethyl cellulose in 100ml ethanol. The resulting phosphor colloid was deposited on the FTO/TiO_2 electrode twice by a repeated doctor blade coating and was treated post-thermally at 500 °C for 0.5 h. The counter electrode was fabricated by dip-coating a FTO glass substrate with a H_2PtCl_6 (Sigma-Aldrich, 37.5% Pt basis) solution (2-mg Pt in 1-ml ethanol), followed by calcining at 400 °C for 0.5 h.

The morphology of the samples was examined by scanning electron microscopy (SEM, Hitachi-S4700). The current-voltage curves of cells were measured under a simulated AM 1.5 G illumination with a light intensity of 100 mW cm^{-2} (300 W, model 91160B, Oriel) with the aid of a computer-controlled potentiostat (CHI-660B, CH Instruments). The active area of the cell was 0.25 cm^2. The incident photon to current conversion efficiencies (IPCEs) were measured as a function of the wavelength from 300 to 800 nm using a solar cell spectral response measurement system (PV Measurements, Inc. QEX7).

3. Results and Discussion

The SEM images of the anodic TNTs on the Ti-mesh substrates are shown in Figure 1. After anodization, the mesh remains sufficient void regions. The TNTs grew perpendicularly to the surface of the titanium wire and covered the entire wire uniformly. The lengths of the TNTs increased with increasing growth time and were 6.1, 9.3, and 11.4 μm after 4, 6, and 8 h anodization, respectively. The inner diameter of the TNTs retained a constant value of approximately 100 nm over various growth times. With increasing TNT length, cleavage was observed at the top part, which was derived from the curvature of the round-shaped titanium wire. The divergently grown TNTs tended to bundle through van der Waals attraction and capillary forces during drying [22]. When assembling a DSSC, the TNTs might be peeled off partially by an external bending force.

Figure 1. Low-magnification SEM images from top view (a) and cross-section (b) of TNTs on Ti-mesh substrate. Top view SEM image of the anodic TNTs (c). Cross-section SEM images of TNTs anodized for 4 (d), 6 (e) and 8 h (f) [23].

The dye-sensitized TNTs with different lengths grown on mesh substrates were sandwiched with the same sized Pt coated mesh to fabricate DSSCs. Figure 2 and Table 1 present the photocurrent-voltage characteristics of the cells and the key factors, respectively. The cells had a constant open circuit voltage around 0.73 V whereas the short circuit current showed an uptrend when the TNTs became longer. The photocurrent increased from 11.74 to 17.48 mA/cm^2 and the efficiency was improved. This might be contributed by the elevated dye loading in a thicker TNT layer. The cell using 11.5 μm long TNTs yielded the highest conversion efficiency of 5.3% among the cells tested. A properly thick layer could preserve the intact TNT layer around the Ti wires during operation. For comparison the efficiency of a DSSC assembled using TNTs grown on Ti-foil as a photoanode and Pt-coated ITO glass as a counter electrode was measured together with the mesh type DSSCs. Although the foil type cell has a TNT layer with a longer length, the efficiency was much lower, due to the low photocurrent. Therefore, the mesh type electrodes developed in the present study could effectively improve the performance of DSSCs. However, the fill factor of the mesh type cell was lower than the typical value of other DSSCs based on TNTs. This phenomenon might be explained by the long diffusion path through the electrolyte for the redox species to reduce the e^--emitted dye molecules [22].

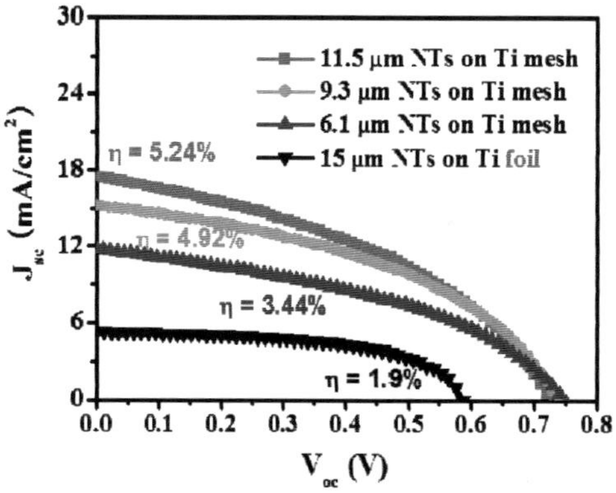

Figure 2. I-V curves of DSSCs utilizing the TNTs with different length on mesh or foil substrates [23].

Table 1. Performances data of the DSSCs based on TNTs/Ti-mesh or TNT/Ti-foils structure [23].

NT Length(µm)	η (%)	V_{oc} (V)	J_{sc} (mA/cm^2)	FF (%)
11.5	5.30	0.72	17.48	42.1
9.3	4.92	0.73	15.21	44.3
6.1	3.67	0.74	11.74	42.2
15 (on foil)	2.24	0.72	5.29	59.0

The enhanced performance of the mesh type DSSCs was attributed to three prominent advantages. First, the photoanode is front-illuminable in the cell, which preserves light harvest from absorption by the counter electrode and electrolyte. In addition, the cell can absorb light from the surroundings at various angles compared to conventional DSSCs based on planar electrodes as both electrodes are the mesh type and the titanium wires of the meshes possesses a circular shape. An efficiency of about 3.15% is obtained by illuminating the back side of the DSSC using 11.5 µm long TNTs. Second, the titanium wires fully covered with TNTs showed a higher surface area than that of TNTs on an equivalent sized foil substrate. Third, after the absorption of solar energy, electrons emitted from the dye molecules should be transported through the TiO_2 layer to the anode substrate and finally to the external circuit.

The standard redox couple used for dye regeneration, iodide/triiodide (I^-/I_3^-), has unmatched performance but typically requires a catalyst during DSSC operation [27]. Although the titanium mesh manifests a good conductivity, it does not meet the demand of assisting the redox reaction between the I^-/I_3^- group. In the present study, a 10 nm thick Pt layer was coated on the Ti-mesh for the counter electrode and a Pt coating can improve the redox reaction rate[NT].

For phosphor particle based DSSC study, the working electrode was prepared in multi-layers. Figures 3(a)~(d) show SEM images of the TiO_2 layer coated FTO glass and the FTO/TiO_2/phosphor electrode. The TiO_2 film contained no aggregates of P25 particles, which were dispersed homogeneously, and formed a porous structure throughout (Figure 3(b)). The thickness of the TiO_2 film was uniform at ~12 μm, as determined by cross-sectional SEM (Figure 3(a)). The phosphor layer had a large degree of roughness due to the nonuniform micro-size of $SrAl_2O_4$:Eu^{2+}, Dy^{3+} particles with a mean thickness of approximately 23.4 μm, as shown in Figures 3(c) and 3(d). The particle size and thickness of TiO_2 SL was approximately 450 nm and 2.1 μm, respectively. The composition of the multi-layer film was confirmed by XRD (data are not shown) and all peaks marked with solid squares were derived from the $SrAl_2O_4$:Eu^{2+}, Dy^{3+} phosphor. The XRD pattern of the TiO_2-coated FTO revealed clear anatase crystallization of the TiO_2 layer.

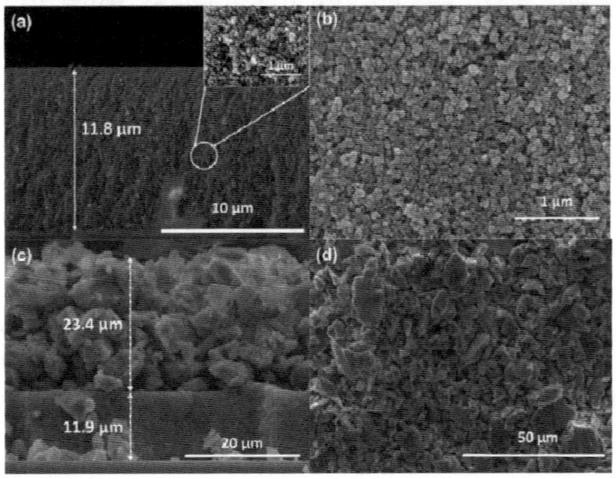

Figure 3. Cross-section and surface SEM images of FTO/TiO_2 (a), (b) and $SrAl_2O_4$:Eu^{2+}, Dy^{3+} FTO/TiO_2/phosphor electrode (c), (d) [24].

$SrAl_2O_4$:Eu^{2+}, Dy^{3+} can be excited by natural sunlight, and its phosphorescence is derived from a transition from the 5d level to the 4f level of Eu^{2+}.[32] Regarding the mechanism for protracted afterglow, the co-doped Dy^{3+} ions are believed to act as hole trap levels, which play an important role in this process[25]. Under UV illumination, the excitation transitions of Eu^{2+} $4f^7 \rightarrow 4f^6 5d^1$ occur immediately, generating a large number of electrons and holes in the vicinity of the valence band (VB). The holes migrate through the valence band and are captured by Dy^{3+} ions located at a suitable depth to form Dy^{4+} ions [JPCC]. The return to the ground state of Eu^{2+} with light emission is triggered by the thermo-activated promotion of an electron from the valance band to the first unoccupied levels of Dy^{4+}, followed by the transition of trapped holes to a photo-generated Eu^+ cation. This process causes the long afterglow of the phosphors.

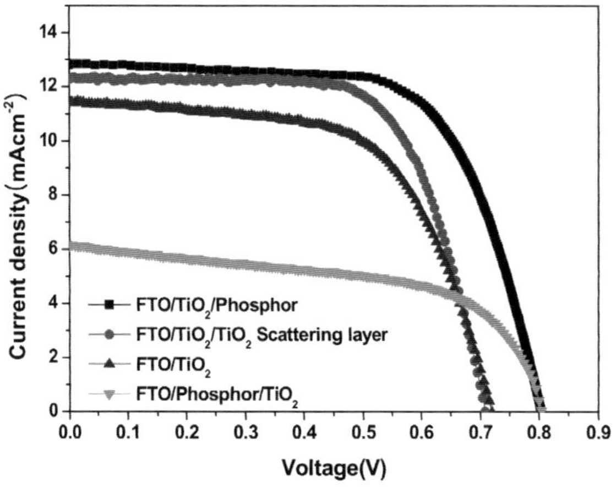

Figure 4. I-V curves of DSSCs using different photoelectrodes: FTO/TiO_2/phosphor (black), FTO/TiO_2/TiO_2 SL (green), FTO/TiO_2 (brown) and FTO/phosphor/TiO_2 (orange) [24].

Table 2. Performances data of the DSSCs based on different photoelectrode structures [24].

Photoanode type	J_{SC} (mA/cm^2)	V_{OC} (V)	FF (%)	η (%)
FTO/TiO_2	11.5	0.72	60.9	5.04
FTO/TiO_2/TSL	12.27	0.71	65.5	5.92
FTO/TiO_2/Phosphor	12.84	0.80	66.4	6.85
FTO/Phosphor/TiO_2	6.14	0.81	56.8	2.84

Figure 4 shows the photovoltaic performances of the phosphor-assisted DSSCs, scattering layer-inserted DSSCs and pure-TiO_2-based DSSC, and Table 2 lists the key parameters. When the phosphor material was inserted into the DSSCs, the open-circuit voltage (V_{OC}) increased to approximately 0.8 V from 0.72 V, which is typical for pure-TiO_2 devices. On the other hand, the short-circuit current density (J_{SC}) increased only when the phosphor material was placed at a proper position in terms of the FTO/TiO_2/phosphor multi-layer structure. The photocurrent of DSSCs with the FTO/phosphor/TiO_2 electrode decreased to 6.14 mA/cm^2. This might be due to mismatch between the electronic band structure of the phosphor and that of the FTO substrate and dyes, so that the electrons cannot be collected effectively and are transported by the phosphor. The presence of a phosphor layer between FTO and TiO_2 also

IOP Conf. Series: Materials Science and Engineering **54** (2014) 012025 doi:10.1088/1757-899X/54/1/012025

increases the length of the electron-transporting path, which would raise the electron/I_3^- recombination rate. The cell with the different layer structure: i.e., FTO/TiO_2/TiO_2 SL electrode showed an obvious increase in photocurrent compared to the pure TiO_2 based DSSC due to the insertion of a scattering layer. Nevertheless, it was still slightly lower than the photocurrent of the FTO/TiO_2/phosphor type DSSCs. The total conversion efficiency of the FTO/TiO_2/TiO_2 SL-based DSSCs was much lower than that of the FTO/TiO_2/phosphor-based DSSCs due to the lower open-circuit voltage. In the device performance, the DSSCs with the multi-layer structure in the order of the FTO/TiO_2/phosphor were found to produce the highest current density of 12.84 mA/cm^2, resulting in the highest conversion efficiency of 6.85%.

The increase in current could be explained by promoted light harvesting. As shown in Figure 5, the IPCE of the phosphor-assisted DSSC are higher in both the UV and visible regions, whereas it was only improved in the visible region in the case of TiO_2 SL-inserted DSSC. An apparent decrease in IPCE in the range from 350 to 400 nm could be observed for the pure-TiO_2 based DSSC and TiO_2 SL-inserted DSSC, and this drop did not appear for the phosphor-assisted DSSCs. The $SrAl_2O_4$:Eu^{2+}, Dy^{3+} phosphor could absorb near UV-photons, giving rise to characteristic visible green emission, which is a unique property that the TiO_2 scattering layer does not possess. Green light was emitted from the excited dye N719 and yielded excess electrons. In this way, solar energy was converted to electric energy more effectively than by the DSSC without phosphors, resulting in an increase in quantum efficiency. In the visible light range, the improved IPCE of both the phosphor-assisted DSSC and TiO_2 SL-inserted DSSC could be explained by the presence of a large-size particle layer on top of the TiO_2 layer, prolonging the optical-path length due to light scattering. Consequently, an increase in IPCE in the full light range could be achieved by adding a phosphor layer on top of the TiO_2 conducting layer.

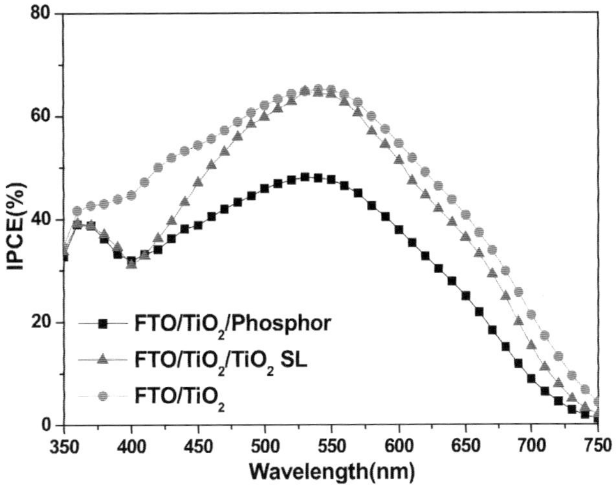

Figure 5. IPCE spectra of pure-TiO_2 DSSC (black squares), TiO_2 SL inserted DSSC (pink triangles) and phosphor assisted DSSC (orange circles) [24].

4. Conclusion

The cell based on a 11.5 μm long TNTs/Ti-mesh structure exhibited the best performance of 5.3%. The mesh type DSSCs has several unique advantages such as availability of front-illumination, high surface area, and low sheet resistance. The influence of the Pt layer on the impedance was also studied. A FTO/TiO_2/phosphor multi-layered photoelectrode was also fabricated for DSSCs by depositing a layer of a long-persistent phosphor, $SrAl_2O_4:Eu^{2+}$, Dy^{3+}, on a porous TiO_2 film. The conversion efficiency of the phosphor-assisted DSSCs was improved to 6.85%, which is more than 13% higher than that of pure-TiO_2-based DSSCs.

References

[1] O'Regan B and Grätzel M, 1991 *Nature* **353** 737
[2] Fan X, Wang F, Chu Z, Chen L, Zhang C, and Zou D, 2007 *Appl. Phys. Lett.* **90** 073501
[3] Wang N, Han L, He H, Park N H, and Koumoto K 2011 *Energy Environ. Sci* **4** 3676
[4] Ding I K, Zhu J, Cai W, Moon S J, Cai N, Wang P, Zakeeruddin S M, Grätzel M, Brongersma M L, Cui Y, and McGehee M D 2011 *Adv. Energy Mater.* **1** 52
[5] Shan G B and Demopoulos G P 2010 *Adv. Mater.* **22** 4373
[6] Seo Y G, Woo K, Kim J, Lee H, and Lee W, 2011 *Adv. Funct. Mater.* **21** 3094
[8] Chang C H and Lee Y L, 2007 *Appl. Phys. Lett.* **91** 053503
[9] Pinilla C, Pópolo M G D, Lynden-Bell R M and Kohanoff J, 2005 *J. Phys. Chem. B* **109** 17922
[10] Xue B, Fu Z, Li H, Liu X, Cheng S, Yao J, Li D, Chen L, and Meng Q, 2006 *J. Am. Chem. Soc.* **128** 8720
[11] Wang Y, Sun Y, Song B, Xi J, 2008 *Sol. Energy Mater. Sol. Cells* **92** 660
[12] Lee C P, Lee K M, Chen P Y, Ho K C, 2009 *Sol. Energy Mater. Sol. Cells* **93** 1411
[13] Qin P, Zhu H, Edvinsson T, Boschloo G, Hagfeldt A and Sun L, 2008 *J. Am. Chem. Soc.* **130** 8570
[14] Mora-Seró I, Fabregat-Santiago F, Denier B, Bisquert J, Tena-Zaera R, Elias J, and Lévy-Clément C, 2006 *Appl. Phys. Lett.* **89** 203117
[15] Snaith H and Schmidt-Mende L, 2007 *Adv. Mater.* **19** 3187
[16] Zwilling V, Darque-Ceretti E, Boutry-Forveille A, David D, Perrin M Y and Aucouturier M, 1999 *Surf. Interface Anal.* **27** 629
[17] Lee S H A, Abrams N M, Hoertz P G, Barber G D, Halaoui L I, Mallouk T E, 2008 *J. Phys. Chem. B* **112** 14415
[18] Nazeeruddin M K, Kay A, Rodicio I, 1993 *J. Am. Chem. Soc.* **115** 6382
[19] Shan G B, Demopoulos G P, 2010 *Adv. Mater.* **22** 4373
[20] Zhu G, Wang X J, L, H L, Pan L K, Sun H C, Liu X J, Lv T, Sun Z, 2012 *Chem. Commun.* **48** 958
[21] Huang X Y, Wang J X, Yu D C, Ye S, Zhang Q Y, Sun X W, 2011 *J. Appl. Phys.* **109** 113526
[22] Rustomji C S, Frandsen C J, Jin S, and Tauber M J, 2010 *J. Phys. Chem. B* **114** 14537
[23] He W, Qiu J, Zhuge F, Li X, Lee J H, Kim Y D, Kim H D, Hwang Y H, 2012 *Nanotechnology* **23** 225602
[24] He W, Timur A, Kim H K, and Hwang Y H, 2013 *J. Phys. Chem. C* **117** 17894
[25] Matsuzawa T, Aoki Y, Takeuchi N, Murayama Y, 1999 *J. Electrochem. Soc.* **85** 149

AUTHOR INDEX